Understanding Place

GIS and Mapping across the Curriculum

Diana Stuart Sinton and Jennifer J. Lund

ESRI PRESS
REDLANDS, CALIFORNIA

ESRI Press, 380 New York Street, Redlands, California 92373-8100

Copyright © 2007 ESRI

All rights reserved. First edition 2007
10 09 08 07   1 2 3 4 5 6 7 8 9 10

Printed in the United States of America

Library of Congress Cataloging-in-Publication Data
Sinton, Diana Stuart, 1966–
    Understanding place : GIS and mapping across the curriculum / Diana Stuart Sinton and Jennifer J. Lund.
      p.   cm.
    Includes index.
    ISBN 1-58948-149-6 (pbk. : alk. paper)
    1. Geographic information systems—Study and teaching (Higher) I. Lund, Jennifer J., 1958– II. Title.
    G70.212.S56   2006
    910'.285—dc22   2006025302

Note: Many case studies discuss maps created by undergraduate students. To improve clarity, some maps were redrawn for this book using the original data.

Ask for ESRI Press titles at your local bookstore or order by calling 1-800-447-9778. You can also shop online at www.esri.com/esripress. Outside the United States, contact your local ESRI distributor.

ESRI Press titles are distributed to the trade by the following:

In North America:
Ingram Publisher Services
Toll-free telephone: (800) 648-3104
Toll-free fax: (800) 838-1149
E-mail: customerservice@ingrampublisherservices.com

In the United Kingdom, Europe, and the Middle East:
Transatlantic Publishers Group Ltd.
Telephone: 44 20 7373 2515
Fax: 44 20 7244 1018
E-mail: richard@tpgltd.co.uk

Cover and interior design by Savitri Brant

Contents

Acknowledgments   vii

Thinking with maps   ix
Jennifer J. Lund and Diana Stuart Sinton

What is GIS? A very brief description for the newly curious   xiii
Diana Stuart Sinton and Jennifer J. Lund

Part 1: Teaching students to think spatially using GIS

1   Critical and creative visual thinking   1
Jennifer J. Lund and Diana Stuart Sinton

2   About that G in GIS   19
Diana Stuart Sinton and Sarah Witham Bednarz

3   Finding narratives of time and space   35
David J. Staley

4   Mapping and quantitative reasoning   49
Jennifer J. Lund

5   Campus–community collaborations: Integrating partnerships, service learning, mapping, and GIS   63
Melissa Kesler Gilbert and John B. Krygier

Part 2: GIS case studies in the curriculum

Expanding the social sciences with mapping and GIS   77
Donald G. Janelle

6   Sociology: Surprise and discovery exploring social diversity   83
John Grady

7   Economics: Exploring spatial patterns from the global to the local   97
James Booker

8   Anthropology: Mapping Guinea savanna ecology in Sierra Leone   113
A. Endre Nyerges with Daniel M. Saman and Laura B. Whitaker

9   Political science: Redistricting for justice and power   129
Mark Rush and John Blackburn

10   Urban studies: Assessing neighborhood change with GIS   141
Christine Drennon

11   Classical archaeology: Building a GIS of the ancient Mediterranean   155
Pedar W. Foss and Rebecca K. Schindler

Teaching the natural sciences with GIS   167
Diana Stuart Sinton

12   Biology: Spatial investigations of populations and landscapes   171
M. Siobhan Fennessy, E. Raymond Heithaus, and Robert A. Mauck

13   Environmental studies: Interdisciplinary research on Maine lakes   187
Philip J. Nyhus, F. Russell Cole, David H. Firmage, Daniel Tierney, Susan W. Cole, Raymond B. Phillips, and Edward H. Yeterian

14   Chemistry and environmental science: Investigating soil erosion and deposition in the lab and field   201
Karl Korfmacher, Brigitte Ramos, and Shelie Miller

15   Geology: Long-term hydrologic impacts of land-use change   213
Suresh Muthukrishnan

GIS and spatial thinking in the arts, humanities, and languages   223
Jennifer J. Lund

16   Foreign language and sociology: Exploring French society and culture   227
Joel Goldfield and Kurt Schlichting

17   Historical geography: Mapping our architectural heritage   237
Robert Summerby-Murray

18   Religious studies: Exploring pluralism and diversity   249
Patrice C. Brodeur and Beverly A. Chomiak

19   Musicology: Mapping music and musicians   259
Jennifer J. Lund

Index   *269*

Topical cross-reference   *284*

Acknowledgments

This volume emerged from work sponsored by NITLE, the National Institute for Technology and Liberal Education, and its former regional hub, the Center for Educational Technology at Middlebury College. We thank the Andrew W. Mellon Foundation for recognizing that mapping and GIS have a role in undergraduate liberal education.

The editors wish to thank all the superb authors and reviewers, particularly Anne Kelly Knowles, who contributed to many aspects of this book. We are grateful to Claudia Naber, Carmen Fye, Michael Law, Colleen Langley, Savitri Brant, Judy Hawkins, and everyone at ESRI Press who made this book possible.

For their support, encouragement, and inspiration during this process, Diana personally thanks Jo Ellen Parker, Amy McGill, Jamie Carroll, and all other NITLE staff, particulary the Vermont-based x-crew; the Advisory Group for Latitude, the NITLE GIS Initiative; Clara Yu; Sarah Bednarz; Bill Hegman; Pete Nelson; Mike Goodchild; Sara Fabrikant; Stuart Sweeney; Don Janelle and other CSISS/SPACE staff; Amy Hillier; Marsha Alibrandi; Jon Caris; and David Rumsey. She sends unending gratitude to Chris and the three little ones, Emily, Eric, and Julia, for forgiving many absent evenings and weekends.

Jennifer thanks her wonderful family of friends and relatives for their patience and support. In particular, she thanks Kathleen Ebert-Zawasky and Jim Schroeder for their daily contributions to the project, both direct and indirect. She thanks her GIS collaborators, the dedicated and innovative Wheaton faculty, especially their visionary provost emeritus, Susanne Woods. She acknowledges the pivotal role of Anne Kelly Knowles who introduced her to GIS at Wellesley College, an encounter that shifted her career two compass points toward a passion. An unseen partner was her mentor, Jim Malanson, contributing many years of wise and patient coaching in courage, calm, and the craft of change.

This book is dedicated to the memory of Bob Churchill, a geographer whose work with mapping and GIS inspired thousands of students and faculty over the decades. May he be mapping in peace.

Thinking with maps

Jennifer J. Lund and Diana Stuart Sinton

Technology seems always on the verge of revolutionizing education. In the 1940s, people thought television would replace teachers. In the 1960s, we thought "programmed instruction" would guarantee learning. The 1980s brought personal computers, gaming, and simulations. In the 1990s, it was the Internet.

GIS mapping software is not the educational revolution of this century's first decade, but it is a powerful tool for teaching and learning. Each case study in this volume describes how an experienced instructor has used GIS in the service of his or her own teaching, within the traditions of a classical undergraduate education. Authors describe how they integrated mapping software into their syllabi as they pursued the learning goals of their discipline and worked to create a realistic learning environment in which students practice inquiry in their field. Chapters span the natural sciences, social sciences, arts, and humanities.

The innovations are tremendously valuable. When students use GIS and mapping to augment their inquiry, they see more, understand more, and engage more deeply with the subject matter. Their efforts bring larger rewards in fewer weeks. They are empowered to pursue their own questions and curiosities. They investigate pressing local issues and make valuable contributions to their communities.

Thinking globally

Our world is one where nations and individuals are becoming more tightly interdependent with passing time and maturing technologies. In the book *Why Geography Matters,* Harm De Blij describes three looming reasons to attend to the "mesh of civilizations": the climate, the economy, and terrorism (De Blij 2005). As we strive to prepare ourselves and our young people for the challenges of the future, we must acknowledge that countless global changes can affect us, our nations, and our world. A contemporary college education must address the topic of geographic relationships—not a familiarity with names and places (although that is important), but an awareness of powerful spatial concepts such as distance, proximity, and pattern recognition.

The multifaceted influence of place should be studied for many reasons. Spatial thinking is as important as logical thinking or quantitative thinking; it is a cognitive skill necessary to understand the ubiquitous influence of location. Looking at a location is an inherently cross-disciplinary undertaking. When trying to understand a place—whether Manhattan, New York or Manhattan, Kansas—we study the ecology, history, politics, economics, sociology, culture, and more. Looking through an interdisciplinary lens builds cognitive muscle, and students learn to consider phenomena through multiple perspectives with realistic complexity and ambiguity. Important thinking skills accrue when studying a place, whether the off-campus neighborhood or an ancient civilization.

Using maps to support the goals of undergraduate education

The first section of this book takes a pedagogical approach. It describes the role of maps and GIS in teaching important skills: thinking with visual evidence, thinking spatially, creating narratives, reasoning with quantities, and collaborating with communities.

In 2004, the Association of American Colleges and Universities (AAC&U) outlined five critical outcomes of higher education (Association of American Colleges and Universities 2004). The first is strong analytical, communication, quantitative, and information skills. We delineate how these critical thinking skills can be conveyed and practiced through class work and assignments using GIS. At Ohio State University, David J. Staley uses maps and GIS to encourage students to reflect on the subtleties of history and to communicate their interpretations to a reader, using a map as vividly as text. In the words of another scholar, "mapmaking, like writing, is a generative act" (MacWilliams 2005). Jennifer J. Lund, describing her experiences at Wheaton College in Massachusetts, proposes that GIS, as a visual calculator, encourages even reluctant students to develop their quantitative skills. Whether analyzing populations or rainfall, students are more eager to undertake quantitative operations when their efforts produce a map. Diana Sinton, drawing on her experiences as a geography professor and at the National Institute for Technology and Liberal Education (NITLE), recounts the importance of spatial thinking in learning how to use complex information. With GIS, students can interact with data visually, managing massive amounts of information and using the data in the service of their intellectual pursuits. At Kenyon College in Ohio, biology students participate in an ongoing research project monitoring the long-term ecological variation of a nearby forest. They progress from initial hypothesis to final analysis using GIS as a tool in their sampling, data management, and analysis.

Another critical outcome emphasized by the AAC&U is for students to understand and experience scholarly inquiry as practiced by the various disciplines. Case studies in this book illustrate how students in the sciences practice scientific inquiry and how students in the humanities practice the methods of their fields. In part II, anthropologist A. Endre Nyerges describes how his students at Centre College conducted spatial analyses to challenge the accepted interpretation of deforestation of the African savannah. The computing power of GIS, coupled with its visual output, lets students participate in more comprehensive scholarly inquiry.

Many of the case studies in this book speak of the rewards that emerge from studying a place with the goal of improving its well-being. The AAC&U endorses an educational experience that instills a sense of responsibility for the world, and responsibility for choices made as an individual and as a citizen. Students in an environmental science course at Maine's Colby College engage in community service analyzing the health of local lakes. They use GIS models to evaluate and predict the health of lakes facing heavy residential development. After sharing their findings with several lake associations, the college now receives requests to study specific lakes that inform local planning and zoning decisions.

Many GIS projects focus on the study of the "other," pursuing the cross-cultural understanding at the heart of a liberal arts education. The religious studies students of Patrice Brodeur surveyed local religious communities and used GIS with census data to document characteristics of the community. They crossed community boundaries on multiple fronts, leaving the campus to interact with residents of the city, often investigating communities whose ethnicity and religious traditions differed from their own.

Another benefit of cross-cultural engagement and community-based service is developing habits of mind that foster integrative thinking. In an urban studies course, Professor Christine Drennon's students at Trinity University responded to a request from the local Habitat for Humanity to assess the impact of their work on the city of San Antonio. Students were challenged by the complexity of the task and the breadth of knowledge required. The interdisciplinary teams worked together to define and complete the study, drawing on their collective knowledge of political science, sociology, and economics to manage the complexity inherent in realistic social research.

In these stories, GIS is a facilitating tool; student learning is focused on the academic content of the course, whether ecology, religion, or urban studies. As a tool, the software empowers students to perform more sophisticated investigations than otherwise possible. GIS facilitates data management, spatial analysis, and visual presentation of results. It enables students to engage strongly with their research process and results, becoming active participants in the inquiry and learning that accrues. The many contributors to this anthology have eagerly shared their successes and candidly discussed the challenges of bringing GIS into the college classroom. Our collective purpose is to help other teachers deepen student engagement by using GIS to manage, analyze, and visualize information.

The first section of the book addresses overall teaching goals, and the chapters that follow describe specific examples. Stories are from all divisions of the academy and testify to the broad relevance of mapping to the academic disciplines. The majority of chapters are presented as case studies. A few chapters survey educational applications of GIS in their field. Chapters are organized by discipline and can be read in any order. The topical cross-reference identifies keywords by educational topic such as "collaboration," and by the nature of data such as "forest." This structure is intended to facilitate sharing ideas between disciplines. As of this writing, relatively few college and university instructors have integrated maps and GIS into their courses. These examples are a tiny subset of possible applications and an indication of the many other creative stories that we hope will follow.

References

Association of American Colleges and Universities (AAC&U). 2004. Liberal Education Outcomes. www.aacu .org/advocacy/pdfs/LEAP_Report_FINAL.pdf/assessment.

De Blij, Harm. 2005. *Why geography matters.* New York: Oxford University Press.

MacWilliams, Mark. Department of Religious Studies, St. Lawrence University. November 5, 2005. Personal communication to Jennifer Lund.

What is GIS? A very brief description for the newly curious

Diana Stuart Sinton and Jennifer J. Lund

Geographic information systems (GIS) combine maps with tables of information. The information—words, numbers, or images—can be linked to locations on the map. For example, a history professor discussing Sir Francis Drake's perilous voyage around the world (1577–1580) might begin with the map in figure 1, showing the segments of his three-year journey (SAGUARO Project 2006).

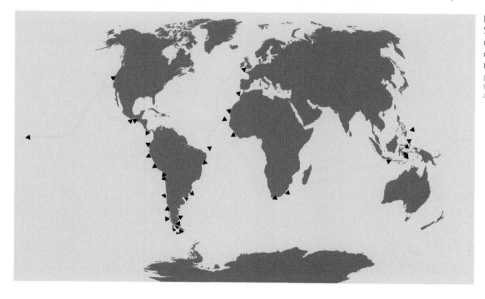

Figure 1.
Sir Frances Drake's circumnavigation of the world (1577–1580).

Data courtesy of Scott Walker, SAGUARO Project, and ESRI Data & Maps 2004.

In figure 2 we have zoomed in to see the tip of South America. The looping path hints at navigational difficulty, and we want to know more. Information about each segment of the journey is held in one row of the table. Because the table is linked to the map, clicking a row in the table highlights that segment of the trip. Conversely, clicking a segment of the route on the map highlights the corresponding row in the table.

To make the description easier to read, we asked the software to display just that single row in a separate window. We see the summary, "Emerging from the Strait, Drake's fleet is driven southward

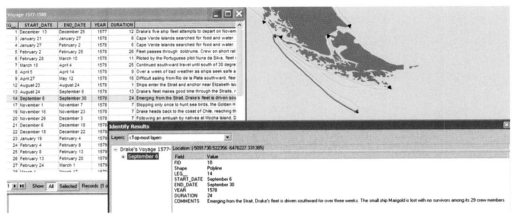

Figure 2. Drake's path near Cape Horn. Drake was sailing westward around the globe. The summary of information from each segment of the journey was extracted from the captain's logs and other historical records.
Data courtesy of Scott Walker, SAGUARO Project, and ESRI Data & Maps 2004.

for over three weeks. The small ship Marigold is lost with no survivors among its twenty-nine crew members." Students reading about the tragedy of this trip are learning about Drake and are also finding drama in data.

In addition to teaching students that maps and tables hold stories waiting to be revealed, GIS can also add context to the narrative by adding information from diverse sources. In figure 3, Drake's voyage has been layered onto a map of wind speeds. We have no wind measurements from the sixteenth century, but this contemporary data shows the conditions that Drake's fleet might have been experiencing.

We gain important insights by looking at data displayed as maps. We can also search and sort that data using GIS to enhance our "vision" via computation. For example, we could search for all trip segments with the words "perished," "lost," or "dead" in the descriptions to focus on the danger of the trip. Students might speculate on lack of food and water, and search for both loss of life and number of days between landings. They could question whether

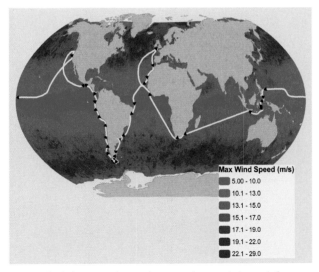

Figure 3. Drake's voyage depicted over maximum wind speeds from 2000 and 2001. Speeds are measured in meters per second.
Data courtesy of Scott Walker, SAGUARO Project, and ESRI Data & Maps 2004

crew members died in hostile encounters with the natives living near the coast. The ability to ask questions makes the data more accessible. The visual output makes it still more accessible. Students' attention is keenly focused on the experiences of the fleet, which animates their efforts to decipher historical records.

The path from a spreadsheet to a mapped representation of data may initially seem mysterious to newcomers, but conceptually it is straightforward. A mapmaker might begin with a blank image of the contiguous United States as a basemap, with each state a separate entity (figure 4). The mapmaker also starts with a table of information, such as the population of each of these states. The GIS software links the "Alabama" population to the "Alabama" shape on the map, the "Arkansas" population to the "Arkansas" shape, and so on for each state. A mapmaker can then find a state's population either by reading the original table or by clicking on the state in the map. This works for any kind of location-based data. A mapmaker can join dots on a map to information about cities, or link lines on a map to information about rivers. Any shape, point, or line can be linked to any kind of information, as long as the feature on the map and the row in the table share a common label, such as the name of the state.

If you want to make a map of your own data, the simplest way is to begin with an existing, GIS-ready basemap. Many sources for such files exist. GIS software companies often supply basemaps with the shapes of countries, cities, oceans, and so on. Thousands of GIS maps are available on the Internet from governments, libraries, commercial distributors, and community-minded researchers and citizens. Alternatively, you can map your own environment by collecting data with global positioning system (GPS) units and then downloading the information into a computer. Scholars can map historic areas by scanning and digitizing antique maps, though copyright laws must be respected.

Once you have a basemap of your area of interest, you can enhance it with data from any number of sources. You can use a table of information from your own research or data gathered from a book or the Internet. For example, you can make a map for something as relatively obscure as the number of deaths caused by falling by downloading data from the U.S. Centers for Disease Control and Prevention (figure 5). (CDC 2005).

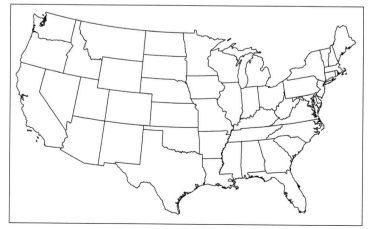

Figure 4. Contiguous United States as a blank basemap.
Courtesy of ESRI Data and Maps 2004.

Location Code	State Name	Death Count	Population	Crude Death Rate
01000	Alabama	159	4,478,403	3.6
02000	Alaska	20	641,616	3.1
04000	Arizona	465	5,441,713	8.5
05000	Arkansas	169	2,706,174	6.2
06000	California	1,470	35,010,990	4.2
08000	Colorado	310	4,502,435	6.9
09000	Connecticut	176	3,459,004	5.1
10000	Delaware	40	806,246	5.0
11000	District of Columbia	40	568,998	7.0
12000	Florida	1,197	16,687,068	7.2
13000	Georgia	441	8,542,720	5.2
15000	Hawaii	70	1,239,971	5.6
16000	Idaho	113	1,343,805	8.4
17000	Illinois	561	12,585,684	4.5
18000	Indiana	254	6,157,782	4.1

Figure 5. Subset of spreadsheet showing number of deaths from falls, by state, in 2002. The Crude Death Rate represents a rate per 100,000 people.

Data from the U.S. Department of Health and Human Services, Centers for Disease Control and Prevention.

After linking the table to the map, you can choose how to display the information. Creating a simple display, such as a map that shows different sizes of symbols based on the total number of falls, is straightforward (figure 6). The message delivered by this map is clear: the greatest numbers of falls happened in California, New York, Texas, and Florida, though this figure does not convey that these states have very large populations overall. Not surprisingly, the largest numbers of falls occur in the states with the largest numbers of people.

By taking into account each state's population, you can create a rate of death from falls, such as the number of falls per 100,000 people in each state (figure 7). In this telling of the tale, California, New York, Texas, and Florida each have small dots, representing low rates of fall-related deaths.

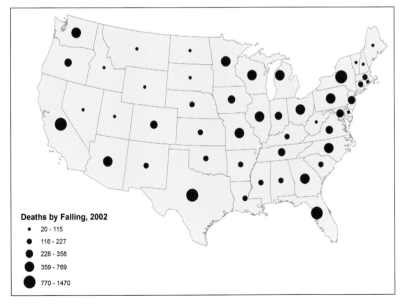

Deaths by Falling, 2002
- 20 - 115
- 116 - 227
- 228 - 358
- 359 - 769
- 770 - 1470

Figure 6. Total number of deaths from falls in each state, 2002.

Data courtesy of ESRI Data & Maps 2004. Additional data from the U.S. Department of Health and Human Services, Centers for Disease Control and Prevention.

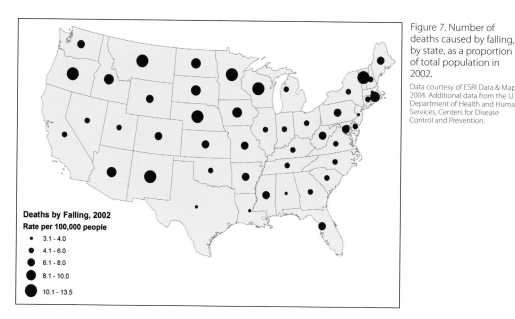

Figure 7. Number of deaths caused by falling, by state, as a proportion of total population in 2002.

Data courtesy of ESRI Data & Maps 2004. Additional data from the U.S. Department of Health and Human Services, Centers for Disease Control and Prevention.

Deaths by Falling, 2002
Rate per 100,000 people
- 3.1 - 4.0
- 4.1 - 6.0
- 6.1 - 8.0
- 8.1 - 10.0
- 10.1 - 13.5

Some less-populated rural states show a contrary pattern. Montana, Nebraska, and New Mexico, with relatively few people, have a much higher rate of death from falling. Observations like these provide opportunities for students to investigate the stories behind the images. They might seek data about the natural world such as icy weather, demographics such as populations of elderly residents, or the built environment such as safety regulations and building codes.

Comparing these maps, we see that mapmaking and map reading are not mechanical activities; they entail discernment, understanding, and communication skill. GIS provides an extensive range of display options, allowing us an extensive range of expression. Our maps can emphasize different aspects of the data to deliver very different messages.

GIS gives users control over what, how, when, and where its information is displayed. It is this control, within an interactive visualization environment, that makes GIS such a valuable teaching and learning environment (Medyckyj-Scott 1994). John Dewey, the father of modern education, put forth the enduring principle that education should be "of, by, and for experience"(Dewey 1938, p. 29). The case studies in this book describe how faculty members bring experiences to their students—letting them participate fully in the adventure of finding meaning and expanding their understanding—using a GIS mapping environment.

References

Centers for Disease Control (CDC). 2005. ICD-10 code (GR099) used for death from all types of falls. wonder.cdc.gov/mortsql.html.

Dewey, J. 1938. *Experience and education.* New York: The Macmillan Company.

Medyckyj-Scott, D. 1994. Visualization and human-computer interaction in GIS. In *Visualization in geographical information systems,* eds. Hearnshaw and Unwin, 200–211. New York: John Wiley & Sons.

SAGUARO Project. 2006. www.scieds.com/saguaro/index.html.

Chapter 1

Critical and creative visual thinking

Jennifer J. Lund and Diana Stuart Sinton

GIS can transform a table of data into a lively artifact to be handled, studied, and challenged. A map transports us to another place or time, to a microscopic or galactic scale. We can look deep within the earth or at the surface of a distant moon. We can trace continental or neighborhood migrations of people or plants across a season or a millennium. We can view manifestations of economies or religions, opinions or disease.

We can engage our students with these maps. We can convey complex and subtle meaning with a map as effectively as we can with a lecture or a chapter of text (Arnheim 1969). Students' capacity for visual complexity can often outpace their cognitive capacity; they can see and remember a complex image long before they achieve an analytical understanding of its meaning. Our distant ancestors lived or died based on their ability to recognize and respond to changing ripples in the grass and shifting colors in the sky (Caine and Caine 1991). Perhaps our students draw on an evolved competency when they perceive and respond to changing patterns on a map.

Yet even with an inborn visual competency, we need education and experience to interpret what we see. A scholar requires a deeper vision to see beyond the obvious and a wiser vision to evaluate what is seen. The American Association of Colleges and Universities (AAC&U) describes critical thinking as an essential capability for a well-educated adult (AAC&U 2004). In our increasingly visual world, we need the ability to read and evaluate images as confidently and thoughtfully as we read and evaluate text. Our visual intuition must be tempered by evidence and critical thought. We should embrace the pursuit of disciplined perception, even while we revel in the joyful leaps of discovery.

This chapter describes how GIS can help students develop the habit of perceiving, evaluating, and analyzing visual information. Interactive maps can help students extend their inborn visual skills with higher-level cognitive skills, creating habits of critical thinking that apply to maps and images as well as to essays and articles.

Illustrating the invisible

Humans seem hardwired to understand spatial relations. The Gestalt school of psychology, formed in Germany in the 1920s and 1930s, documented the meanings we intuitively draw from various spatial relationships such as proximity, similarity, and relative size (Ellis 1938; MacEachren 1995). Using those hardwired perceptions, we can communicate meaning by showing the relationships between objects (Bang 2000) (figure 1).

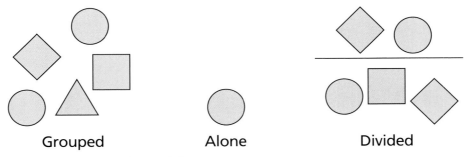

Figure 1. Meaning attributed to spatial relations.
After Bang 2000.

When data is displayed as shapes and lines, those shapes and lines become building blocks of understanding. Maps let us see relationships between visible entities like roads and between invisible entities like opinions. Students readily internalize relationships within and between those entities, grouping and categorizing the parts of the whole, deriving meaning from the maps. As one undergraduate GIS user marveled after identifying a pattern in his data, "It isn't something you would see until you actually saw it."

Maps give us the ability to see where invisible items like attitudes, opinions, and relationships between entities are located. A senior at Smith College asked other students how they felt about various places on the campus, such as where they felt safe and where they liked to spend time. When she mapped the results of her interviews, "hot spots" revealed general attitudes toward different locations around the college (figures 2a and 2b).

In another example, in the final class meeting of a senior seminar on Italian politics, a professor brought his students to the GIS lab and set them loose with a map of Italy (Vogler 2005). Having spent several weeks studying the ebbs, flows, and influences of political fortunes, students found the data to be a rich source of events and trends they had studied in depth. They began with a

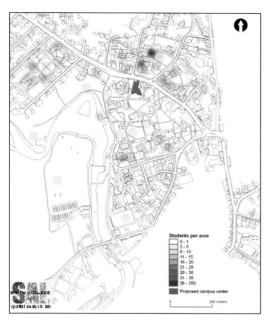

Figure 2a. Where respondents hang out with friends.

Stephanie Keep, "Psychology of Space: Student Perception of Smith College Campus." Courtesy of Smith College Spatial Analysis Lab.

Figure 2b. Where respondents feel least safe.

Stephanie Keep, "Psychology of Space: Student Perception of the Smith College campus." Courtesy of Smith College Spatial Analysis Lab.

data layer displaying the cities with the highest quality of life as reported by a major Italian newspaper between 1988 and 2004 (*Il Sole 24 Ore* 2004). Another layer depicted the voting results of several major elections through the twentieth century. Pairs of students were asked to look at the data for a half hour and present an observation about quality of life and its relationship to voting behavior. One student team used the maps to illustrate its observation that many of the highest scores belonged to medium-sized cities in northern Italy. A second team presented maps of electoral results of the conservative Christian Democratic party (and their successors, the Forza Italia party), showing a persistent strength in the north and south. They used the maps to support the theory that northern and southern Italy have embraced democratic principles more strongly than has central Italy.

This exercise brought polls and elections into the visible realm. Just as a geologist studying a meteor benefits from viewing at least a facsimile representation of the object, these students flourished when given a map to examine spatial data. Traditional academic approaches are enhanced by maps, and maps are necessary to fully understand spatial data (Hallisey 2005).

Making assumptions from maps

One pair of students in the Italian politics class unintentionally demonstrated the powerful human drive to categorize (De Bono 1990) when they began to refer to the smaller symbols as "the less appealing cities." Their peers corrected them, reminding them that the smaller dots represented cities on the list of "10 most desirable cities," albeit at the bottom of the list. Despite our natural visual competencies, we still need to develop an awareness of map legends, and the habit of questioning them, to effectively see and interpret the patterns we perceive in maps. The students' misinterpretation of the small symbols in the quality-of-life map exemplifies a common error of naïve map readers, illustrating the need for careful, not casual, map reading.

The accessible nature of maps prompts viewers to leap to conclusions, as with the 2004 U.S. presidential election map in figure 3, which gives the impression of a landslide victory for George W. Bush, when in fact he received only 51 percent of the popular vote (*New York Times* 2004). This map can fool our well-developed "competency of the eye" (Goffman 1979 p.25). People are likely to accept their raw perceptions and fail to probe an image. The press used this map to display voting results during the election. It is not wrong, but it can be misleading. The map's salient feature is area, and this map is screaming "Red!" Yet land doesn't vote; people do. This map does not tell us that many of the blue states are more densely populated than many of the red states, so it cannot tell us how many votes were received by each candidate. Critical map readers challenge the initial perception of a landslide victory and acknowledge that this map does not deliver the full story on its own.

A map can mislead in many ways. The book *How to Lie with Maps* (Monmonier 1996) enumerates ways to fool an incautious reader, both intentionally and unintentionally, at both the perceptual and the cognitive levels. An innocent map can delude a reader's visual intuition, like the

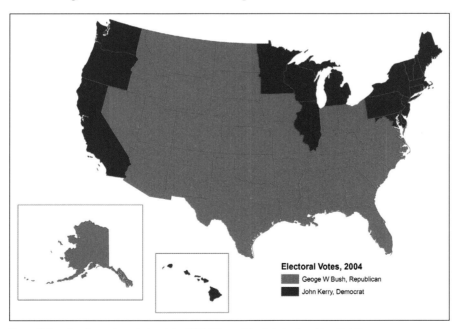

Figure 3. Results of popular vote by state, 2004 U.S. presidential election. *New York Times.*
Data courtesy of Dave Leip (www.uselectionatlas.org) and ESRI Data & Maps.

Figure 4a. Jonathan Carver, 1781, New Map of North America.

Courtesy of David Rumsey Map Collection (www.davidrumsey.com).

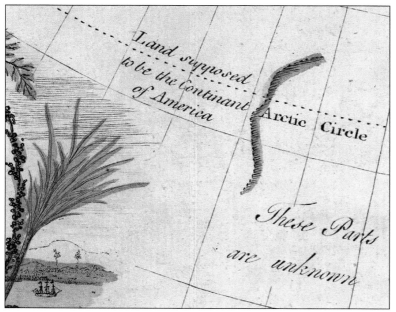

Figure 4b. Close-up from the Carver map, showing text proclaiming "These Parts are unknown."

Courtesy of David Rumsey Map Collection (www.davidrumsey.com).

election map just described. And a mapmaker's point of view can deceive an unsuspecting reader. One eighteenth-century map of North America boldly declares "These Parts are unknown," the mapmaker's shorthand to indicate that no Europeans had explored that region. Innocent of malice, but acting within his social context, he did not acknowledge that the region was well known to the people who inhabited it (figures 4a and 4b).

The practice of critical thinking instills a habit of questioning our initial response to any input. Just as we learn to question an author's point of view, we must learn to question a mapmaker's point of view (Kaiser and Wood 2001). With text we can learn to look beyond incendiary rhetoric and passionate expression. Instead, we can acknowledge our responses to those cues, then evaluate a message with our reason and knowledge, as well as our feelings. With maps we can apply the same process, recognizing biases and agendas of both the map readers and the mapmakers.

GIS can transform undergraduates into mapmakers, empowering them to represent the world as they see it. As mapmakers they are responsible for the implicit messages delivered by their own maps. When students share their maps with classmates, they often illustrate for each other how a map can fail to communicate its intended meaning. As experienced mapmakers, they become more astute map readers, aware that they see a message already translated through the senses and convictions of "a legion of mapmakers, bewildering in their variety" (Wood and Fels 1992).

GIS helps us learn to look beyond the lines and colors, to evaluate a map with reason and method, and to ask questions that could illuminate its message. Why was a particular map made? Who funded its production? What benefit accrued to the mapmaker for emphasizing some elements and omitting others? One teacher observed that the most important aspect of teaching with GIS is "enabling students to learn how to see what isn't there" (Grady 2005).

Changing points of view

In a first-year seminar, students spent an hour with a GIS map of Rhode Island showing the prevalence of autism in school districts, along with several socioeconomic factors (Baron 2005) (figures 5a–5c). Students immediately noticed that wealthier districts had higher proportions of autistic students. One student expressed surprise that families in wealthier areas would be more likely to have autistic children and wondered if some childhood privilege might trigger the syndrome, or if some deprivation offered protection. Well-schooled in critical thinking, the other students immediately began to offer alternative theories. They proposed that wealthier schools may conduct more sensitive diagnoses or that wealthier districts may attract families demanding better services.

When the students noticed this apparent correlation, the stage was set for an important conversation about causality. Visual correlations are powerfully persuasive, and people often leap to closure, asking how one factor caused the other. A teacher can guide that powerful response into a useful line of inquiry. Critical thinkers understand that a visual correlation between autism and wealth

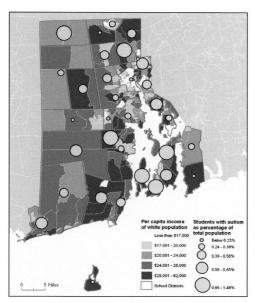

Figure 5a. Autism in Rhode Island. Autism prevalence in Rhode Island (2003–04 from RI Department of Education) displayed over per capita household income by town (2000 from U.S. Census Bureau).

Autism data courtesy of State of Rhode Island and Providence Plantations Department of Elementary and Secondary Education. Additional data from the 2000 U.S. Census.

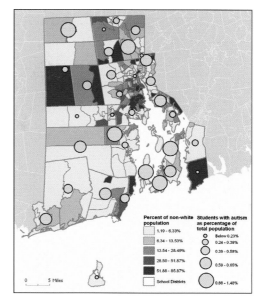

Figure 5b. Percentage of nonwhite residents by town in Rhode Island (2000 from U.S. Census Bureau) and percentage of students with autism by school district (2003–04 from RI Department of Education).

Autism data courtesy of State of Rhode Island and Providence Plantations Department of Elementary and Secondary Education. Additional data from 2000 U.S. Census.

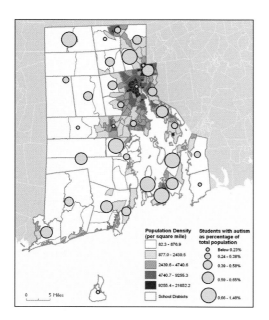

Figure 5c. People per square mile in Rhode Island (2000 from U.S. Census Bureau) and percentage of students with autism by school district (2003–04 from RI Department of Education).

Autism data courtesy of State of Rhode Island and Providence Plantations Department of Elementary and Secondary Education. Additional data from 2000 U.S. Census.

may flag an interesting relationship meriting further investigation, but that demands exploration of all possibilities, such as

- wealth causes autism
- autism causes wealth
- autism and wealth are both related to a third factor, like school services
- autism and wealth are not related, and the visual correlation is coincidental

Students need to internalize the twin messages of "this deserves further investigation" and "correlation does not mean causality." Seeing critically means noticing patterns that raise questions.

As students spend time analyzing relationships between multiple variables, like disease, geography, race, and socioeconomic strata, they develop skills for thinking about multifaceted issues. As they consider first one layer of data, then another, they are learning to don and doff points of view. As they layer or juxtapose data, they learn to see relationships of many kinds. The psychology instructor, leading students in the study of autism, used various data to inspire questions and curiosity about the interplay between factors in the GIS map.

Questioning social assumptions

Examples in this chapter have shown how visual perception must be tempered with careful reading, rational questioning, and prior knowledge. After questioning how our perceptions influence our understanding, we must question how our social assumptions influence our perceptions. Scientists who study vision and cognition agree on relatively few things, but most agree that visual comprehension results from collaboration between the eyes and the brain. Input from the eye is perceived and interpreted based on a person's prior knowledge and expectations. That prior knowledge influences what our eyes see and what our minds comprehend (Olson and Bialystok 1983). In short, we see what we expect to see. It has been shown that researchers are more likely to interpret visual evidence as supporting their own beliefs (MacEachren 1995). Our classroom observations indicate this holds for students reading maps as well.

In one sociology class, while discussing a map of income for eastern Massachusetts residents, the students and the instructor persistently overlooked an unusual map legend (Grady 2004). At one point, an observant student noticed that the highest per capita income for black residents was over a half-million dollars, over five times that of white residents (figures 6a and 6b). The class urgently wanted to know more and immediately located an anomalous tract in Quincy, a neighborhood on the Boston waterfront where the per capita income for black residents was $576,118 (figure 7). This piece of data had escaped their observation through multiple discussions. However, once they became aware of it, they pursued it relentlessly, eager to close the gap between this new information and their previous understanding of a world with a white-dominated economy.

With their assumptions challenged by the data, the students peppered the instructor with questions about the accuracy of the data and the definition of per capita income. The instructor confirmed that for black residents living within that census tract, the median income for a family of four was over two million dollars. As they examined this anomaly, they saw more: the highest

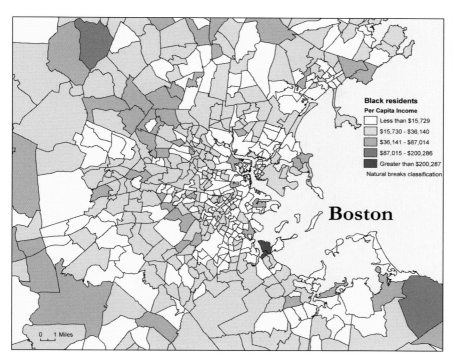

Figure 6a. Per capita income of black residents in eastern Massachusetts, by census tract.

Data from 2000 U.S. Census.

Black residents
Per Capita Income

Less than $15,729
$15,730 - $36,140
$36,141 - $87,014
$87,015 - $200,286
Greater than $200,287

Natural breaks classification

Boston

0 — 1 Miles

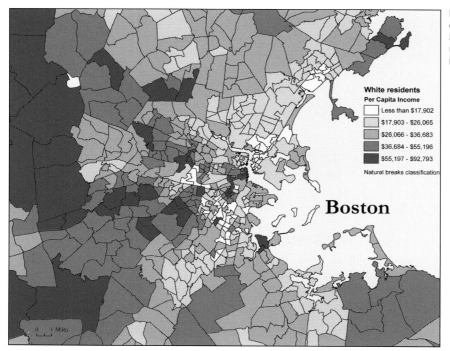

Figure 6b. Per capita income of census tracts in eastern Massachusetts for white residents.

Data from 2000 U.S. Census.

White residents
Per Capita Income

Less than $17,902
$17,903 - $26,065
$26,066 - $36,683
$36,684 - $55,196
$55,197 - $92,793

Natural breaks classification

Boston

0 — 1 Miles

Chapter 1   Critical and creative visual thinking

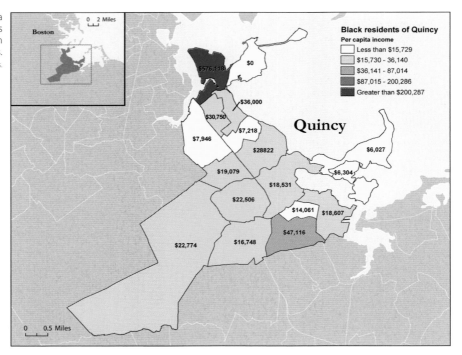

Figure 7. Per capita income of census tracts for black residents in Quincy, Massachusetts.
Data from 2000 U.S. Census.

income bracket for Hispanic residents in eastern Massachusetts was more than three times that of whites and nearly double for Asians. The students' curiosity fueled intense scrutiny of the data, the creation of dozens of maps, and active hypothesizing on social, economic, and arithmetic dimensions. These students held a misapplied assumption that white people consistently earn more than minorities. In reality, minorities living in predominantly white neighborhoods in the U.S. often earn more than their white neighbors (U.S. Census Bureau 2005).

The instructor then introduced the topic of group size and the role that it plays in influencing statistical values. Quincy is a predominantly white, working-class area. Census data shows only a few dozen black residents in that tract. A few extremely wealthy individuals could dramatically shift the median income of that small group. An equal number of comparatively wealthy whites would barely budge the median income of the few thousand white residents.

When the observant student flagged the surprising legend, he flagged the disconnection between his own assumptions and the data. He made the class aware of that implicit assumption, which they seemed to share. The fast pace of asking questions and getting answers made it easier for students to modify their assumptions.

The students were satisfied with the reasoning but still wanted to know why incredibly rich people of any group would choose to live in Quincy. Their hypotheses included doctors, real estate moguls, and entertainment personalities. Using logic and maps of local landmarks, the class concluded that these very wealthy people were probably sports stars because Quincy has excellent access to Boston's sports arenas. Fortuitously, the next day the Boston newspaper featured an article about a waterfront condominium complex in Quincy that was home to several well-known athletes.

Many college instructors comment on how eagerly students engage with GIS maps. Perhaps students' unusually high motivation comes from the maps' visual appeal or perhaps from the interactive game-like nature of the software. Or perhaps their interest springs from a deeper place, the need to find meaning and resolve mysteries. People are driven to make sense of their world, as they are driven to eat and drink. Caine and Caine, in *Making Connections: Teaching and the Human Brain,* state "the search for meaning is innate" and "the search for meaning occurs through 'patterning'" (Caine and Caine 1991).

The drive to resolve, to find meaning, to make sense of the world, kept the sociology students from their lunch as they grappled with the anomaly of a preposterously high median income. In the Quincy example, the picture came into focus for the students only after prolonged scrutiny of the census maps, after learning a new way to look at census information (using group size) and after casting widely for relevant environmental factors (like proximity to sports arenas). This was a memorable classroom episode, yet the enthusiasm is not unusual. Students engage in energetic investigations, push their own limits, and strive to explain a pattern or anomaly. They begin with a map of relevant data and, as with other creative undertakings, "a broad sense of what [they] are looking for, some strategy for how to find it, and an overriding willingness to embrace mistakes and surprises along the way" (Bayles and Orland 2001).

Using evidence to reason

A history seminar had an opportunity to examine census data from Victorian London and these students engaged in their investigation with equal ardor (Crosby 2005). This census data, unlike modern U.S. census data, included personal details. The class had the name, age, birthplace, occupation, and other details of every resident of the Whitechapel district in 1891 (U.K. Data Archive 2002). The instructor used the map to give his students a personal encounter with the place and people of the time. Students had no structured assignment. The instructor assigned each student a portion of the district, entreated them to get to know their neighborhoods, and asked them to share their discoveries with the class the following week.

Among their many observations, students found that immigrants from Eastern Europe tended to live in proximity to one another. The students discovered that silk makers clustered together, shown as yellow dots in figure 8, but butchers were evenly distributed across the district, denoted as green dots in figure 8. The students were highly motivated to find and report generalizations about the populace: the more sweeping their generalization, the more animated they appeared. When they discover patterns, students long for stories to account for what they see, and they begin to discuss, if not debate, how these stories mesh, focusing not on themselves and their opinions, but on the maps and data to be explored.

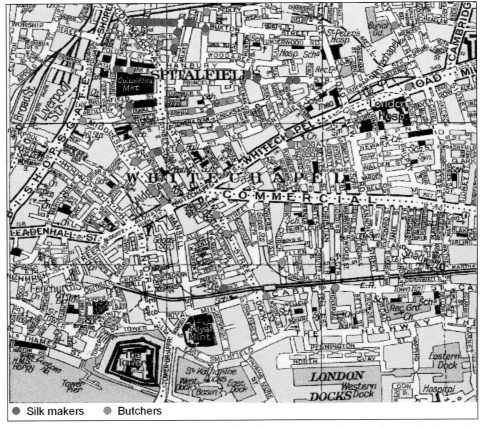

● Silk makers ● Butchers

Figure 8. Charles Booth map of Whitechapel district of London with census information. Points taken from SN 3139 1891 Census Project, Spitalfields.

Data from D. Rau, 1891 Census Project, "Spitalfields," Colchester, Essex, UK Data Archive [distributor], April 1994, SN:3139. Map from the Library of the London School of Economies and Political Science Maps as it appeared in Descriptive of London Poverty. 1898–9. East Central.

These students were using maps as a source of evidence, as envisioned by Jacques Bertin and Edward Tufte, both important thinkers who advocate for graphics as a tool for reasoning. Tufte describes "evidence-based thinking" and advocates "using graphics not just to present a conclusion but to reach it" (Zachry and Thralls 2004). Bertin advocated "using vision to think," creating visual representations continuously as a vehicle for understanding the data and refining the research. He maintained "graphics is the visual means of resolving logical problems" (Bertin 1983).

In the Victorian history class, students generated map after map, seeking illumination, calling out to their classmates when they produced a map with a pattern or visual correlation. Using GIS as a lens, they examined different aspects of their data at various scales. The maps became stepping stones to reach new understanding. The maps provided interim points essential to the conclusion. Students experienced the process of integrating data, context, and reason, an important lesson in the liberal arts curriculum.

Students were engaged as creative historians, no longer studying the maps of others, but inventing and arranging their own map and data compositions. As such, they were at the edge of the known, pushing into the unknown. Their minds were doing the work of "both artist and scientist, attempting to discern and understand patterns as they occur and giving expression to unique and relative patterns of their own" (Hart and Kerslake 1983). Agnes de Mille wrote, "The artist never entirely knows. We guess. We may be wrong, but we take leap after leap in the dark." As these students guessed, they leaped in many directions, following their intuition, generating many maps in search of a pattern or an explanation.

The interactive nature of GIS enables this rapid-fire visualization. Taking a creative, somewhat anarchic path enables serendipitous discovery through observation and inductive reasoning, fostering a close relationship between the data and the mapmaker (in this case, the student) (Sibley 1998). "Proof" and deductive analysis would call for more quantitative tools and techniques, but these descriptive maps fulfilled the objectives of the history professor: for students to engage with the material and to experience the practice of scholarly inquiry.

Practicing to know

As with any creative endeavor, practice is essential to success. Practice helps students move from technical knowing to ingrained, grounded knowing (Oliver 1990). When the practice becomes habit, their minds are freed for more challenging intellectual encounters. An apocryphal story describes an art instructor who announced he would grade half his class according to the quantity of their work. The other half would be graded solely on quality; they needed to produce just one perfect pot. At the end of the term the instructor used a scale to grade the "quantity" students. Students who had produced fifty pounds of pottery received an A, forty pounds was a B, and so on. Surprisingly, the best work came from the students graded on quantity; the "quality" students had spent most of their time reflecting on the nature of a perfect pot and accomplished very little (Bayles and Orland 2001).

The GIS mapping environment lets students achieve that kind of quantity when creating maps. It provides a fertile learning environment in which undergraduates can produce a new knowledge representation every few minutes, immersing themselves in the creative act. They learn from each trial, expanding their understanding through re-ordering and re-presenting the material in various ways. The maps themselves deliver immediate visual feedback on the effectiveness of each effort.

As students create many maps in the span of an hour or a week or a term, they come to appreciate the interlocking dependencies of factors in a complex system. In the autism example, they reflected on possible links between a developmental disease, social context, and physical environment. Even in this simple example, when students merely displayed different combinations of data, they appeared to gain confidence working with multiple variables. This confidence is necessary before they can consider complex issues, entailing multiple perspectives and competing interests.

Global warming is one such issue. When experts debate the impact of the problems and potential remedies, they advocate different points of view. Biologists call for cleaner fuels to reduce pulmonary disease. Economists warn that cleaner fuels would lead to higher prices. Political scientists credit cheap energy with third-world progress. Meteorologists suggest a link between ocean temperatures and catastrophic hurricanes that wreak disproportionate damage in developing nations. Through GIS, students can deconstruct these complex issues into their individual elements, consider those elements individually, and then examine the relationships between them to gain an understanding of the complex whole.

Contemporary institutions of higher education seek to graduate nimble thinkers and responsible citizens, people with the ability to consider diverse data sources and diverse perspectives. The ability to shift one's frame of reference, to peer into a situation through different windows, allows us to shift perspectives in search of insight (Olson and Bialystok 1983). Moving between multiple layers in a GIS map, students discover that their frame of reference is indeed plastic; they can choose their viewpoint, their state of awareness, and what they see. When students develop confidence in altering their frame of reference, they are empowered to question it, to adopt new perspectives, to challenge another person's position, or to seek a better vantage point for a more creative solution. Learning to integrate multiple data sources, multiple perspectives, and multiple ways of thinking is at the core of the liberal arts curriculum.

Creatively making and reading maps

Maps lie at an interesting intersection of art and science. GIS supports both intuitive perception and deductive analysis. The ability to quickly plumb the data and create impromptu maps lets students cast widely for patterns in their search for new meaning. They benefit from the encounter with the unknown and the use of information to prompt insight (De Bono 1990). The case studies in this volume describe the experiences of many educators who have adopted and adapted the application to meet their particular teaching goals.

In college, students learn to explore new ways of thinking and new ways of knowing. Using GIS, they are liberated to be "provocative and permissive: putting information together in new ways and allowing unjustified arrangements of information," breaking up the packaged units of perspectives and premises, of themselves and of others (De Bono 1990). By interacting with the layers of a GIS map, users comb apart the different data sources and the different data types. Looking at each layer individually or in various combinations, they are changing their viewpoint, experimenting with different emphases, pursuing a holistic perspective. They are working to escape the ingrained assumptions of their individual experiences and the blinders of society's assumptions, challenging the information that is displayed and questioning the omissions. As mapmakers and map readers, critical thinkers strive to avoid errors analogous to depicting the sixteenth-century Americas as uninhabited. We want to see and interpret our world effectively, in both cultural and scientific domains.

Humans are pattern-finding creatures, constantly aware of visual repetitions, correlations, and anomalies. This is a deeply rooted strength that drives many opportunities for insight. Yet we must learn to see deeply to extract meaning, to apply context and reason to our visual precepts. We must apply critical thinking practices to balance our tendency to find visual confirmation of what we already believe or have been told. We must learn to look for the unexpected, testing different interpretations to evaluate the merit in alternative readings of patterns we discern.

Today's world is thoroughly multimodal. Our graduates will deal with all manner of communication in their professional lives. Our academic curricula have traditionally emphasized words, but the time has come to embrace the need to think and communicate with images. With GIS, students have the opportunity to develop the skills and habits of seeing and thinking with information-rich images, to become intelligent consumers of visual information and effective communicators with maps.

References

Association of American Colleges and Universities (AAC&U). 2004. Liberal Education Outcomes. www.aacu.org/advocacy/pdfs/LEAP_Report_FINAL.pdf.

Arnheim, Rudolf. 1969. *Visual thinking.* Berkeley, California: University of California Press.

Bang, Molly. 2000. *Picture this: How pictures work.* New York: SeaStar Books.

Baron, Grace. 2005. Visualizing autism. First year seminar. Department of Psychology, Wheaton College.

Bayles, David, and T. Orland. 2001. *Art and fear: Observations on the perils (and rewards) of artmaking.* 1st Image Continuum Press ed. Santa Cruz, Calif.: Image Continuum.

Bertin, Jacques. 1983. *Semiology of graphics.* Madison, Wisc.: University of Wisconsin Press.

Caine, Renate Nummela, and Geoffrey Caine. 1991. *Making connections: Teaching and the human brain.* Alexandria, Va.: Association for Supervision and Curriculum Development.

Crosby, Travis. 2005. Murder and mayhem in Victorian England. Course. Department of History, Wheaton College.

De Bono, Edward. 1990. *Lateral thinking: Creativity step by step.* 1st Perennial ed. New York: Perennial Library.

Ellis, Willis Davis. 1938. *A sourcebook of gestalt psychology.* London: Routledge & Kegan Paul.

Facione, Peter A. 1998. *Critical thinking: What it is and why it counts.* Millbrae, Calif.: California Academic Press.

Goffman, Erving. 1979. Gender Advertisements. London: The MacMillan Press.

Grady, John. 2004. Mapping minorities in the United States Course. Department of Sociology, Wheaton College.

———. 2005. Mobilizing the imagination with visual data. Unpublished manuscript.

Hallisey, Elaine J. 2005. Cartographic visualization: An assessment and epistemological review. *The Professional Geographer* 57(3): 350–64.

Hart, Kathleen, and Daphne Kerslake. April 1983. *Avoidance of fractions.* Paper presented at the annual meeting of the American Educational Research Association. Montreal, Canada.

Il Sole 24 Ore. 2004. Bologna Al Top Della Vivibilità. sec. Qualita' Della Vita.

Kaiser, Ward L., and Denis Wood. 2001. *Seeing through maps: The power of images to shape our world view.* Amherst, Mass.: ODT Inc.

MacEachren, Alan M. 1995. *How maps work: Representation, visualization, and design.* New York: Guilford Press.

Monmonier, Mark S. 1996. *How to lie with maps.* 2nd ed. Chicago: University of Chicago Press.

The New York Times. 2004. Election 2004. www.nytimes.com/ref/elections2004/2004President.html.

Oliver, Donald. 1990. Grounded knowing: A postmodern perspective on teaching and learning. *Educational leadership* 48(1): 64–69.

Olson, David R., and Ellen Bialystok. 1983. *Spatial cognition: The structure and development of mental representations of spatial relations.* Child psychology. Hillsdale, N.J.: L. Erlbaum Associates.

Sibley, D. 1998. Sensations and spatial science: Gratification and anxiety in the production of ordered landscapes. *Environment and Planning A* 30:235–46.

U.K. Data Archive. www.data-archive.ac.uk.

U.S. Census Bureau. 2005. American FactFinder datasets. factfinder.census.gov.

Vogler, David. 2005. Italian politics. Seminar. Department of Political Science, Wheaton College.

Wood, Denis, and John Fels. 1992. *The power of maps.* New York: Guilford Press.

Zachry, Mark, and Charlotte Thralls. An interview with Edward R. Tufte. *Technical Communication Quarterly* 13(4): 447–15.

About the authors

Jennifer J. Lund is the faculty technology liaison for the social sciences at Wheaton College in Norton, Massachusetts. She collaborates with the faculty to design class sessions and to teach students to use GIS for quantitative and spatial reasoning. Lund pioneered small-scale GIS assignments and integrated mapping into diverse courses across the college curriculum. Her passion for maps springs from her observation that students learn math from experience and observation as well as through logic and symbols. She earned a Master of Education degree from Harvard and has many years of experience in academic and corporate education.

Diana Stuart Sinton is the chief program officer for Latitude, a collaborative GIS initiative at NITLE, the National Institute for Technology and Liberal Education (www.nitle.org). NITLE works with its participating colleges to promote the effective use of technology and catalyze innovative teaching. NITLE is an initiative of Ithaka (www.ithaka.org) and receives generous funding from The Andrew W. Mellon Foundation. Sinton holds BA, MS, and PhD degrees from Middlebury College and Oregon State University. She has taught landscape ecology, geography, and GIS at the University of Rhode Island and Alfred University, and has conducted research in the Pacific Northwest and Argentina. Currently she teaches, presents, and writes about the role of GIS in higher education, and takes time to make maps on occasion.

Chapter 2

About that G in GIS

Diana Stuart Sinton and Sarah Witham Bednarz

Students in pursuit of a well-rounded education must learn to think spatially (National Academies Press 2005). Spatial thinking enables us to comprehend and address issues of spatial relationships. As students explore geographical space, they gain the facility and confidence to grasp and imagine abstract spaces, to solve multi-faceted problems, and to think critically and participate actively in our complex, multidimensional world. As a point of entry into the world of spatial thinking, maps, mapping, and GIS have the potential to transform the learning process for students and faculty alike. GIS is a powerful tool that enables students to manipulate and envision multifaceted systems, to use information from many sources, and to consider complex, interconnected problems.

This chapter describes how maps, mapping, and GIS can help students become more spatially aware and appreciate the impact that geographical location has to influence just about everything. We do this by illustrating the connections among spatial thinking, maps, mapping, GIS, critical thinking, and problem solving at a range of scales, from the local to the global, drawing freely from research by geographers, psychologists, and educators who have just begun to articulate the connections among geographic thinking, spatial cognition, and mapping (National Acadamies Press 2005).

The importance of global awareness in the twenty-first century

We keep a great deal of geographic information, gained through experiences and filtered by our perceptions, loosely organized in our brains as "frames of reference" or cognitive maps, the "encoded structure in our long-term memory of what is where" (Lloyd 1997, p.19).

Cognitive maps serve as coat hangers for assorted memories. They provide a vehicle for recall—an image of "where" brings back a recollection of "who" and "what." This sense of place is essential to any ordering of our lives (Downs and Stea 1977, p. 27).

The cognitive maps we maintain most strongly are local ones, obviously because these are the places with which we are most familiar and where we spend most of our time. We remember specific places (the Starbucks, the dry cleaner), can characterize local neighborhoods or nearby areas (the hilly part behind the school, the ritzy section, the congested street), and can use these maps to navigate (how to drive from Starbucks to the dry cleaner without using the traffic-filled street).

Beyond the local or regional scale, most people have highly individual, and often erroneous, understandings of geographic space. Though we may be constantly exposed to a variety of maps, many of us fail to absorb the information they present or to regard them critically. We find it particularly difficult to maintain numerous and accurate cognitive maps of places elsewhere in our country and the world. Collectively, our knowledge of national and global geography is abysmal. Survey after survey finds high percentages of high school and college students who cannot locate their own state on a map, much less the Pacific Ocean or Mexico.[1] Does it really matter that someone doesn't know or confuses their cities? Sometimes, yes. One of the many frightening stories that emerged in the Hurricane Katrina aftermath was about the airplane full of New Orleans evacuees that the Federal Emergency Management Agency (FEMA) inadvertently flew to Charleston, West Virginia, instead of their intended destination: Charleston, South Carolina.[2]

Following natural disasters, such as hurricanes or the 2004 Southeast Asian tsunami, there is a spike in demand for access to maps and imagery, and the growth of online mapping applications has been sudden and rapid.[3] However, despite the availability of aerial photographs and satellite images of nearly any place on earth, the most popular location for which people search is their own house. "Look, I can see my driveway!"

While the novelty of house-finding is a great place to start, these mapping tools are also quickly becoming sophisticated and innovative teaching and learning tools, and the timing couldn't be better. As the world becomes more connected, the need for geographic knowledge becomes greater, and we are faced every day with the fact that geography matters (de Blij 2005). Military strategists recognize that "geography is destiny." Throughout the world of retail business and real estate, "location, location, location" is a mantra. Being able to locate a city on a map, however, is not the primary objective of geographical knowledge. What matters more is understanding the integrated effects of a place's geography: how different physical and human components of environments are interrelated and affect one another. Hurricane Katrina and its impact on the Gulf Coast provide a rich example of this.

We start with a quintessential geography question: Why is it like this here? When Hurricane Katrina struck in late August 2005, several of the levees that normally kept Lake Pontchartrain and its associated canals inside their banks were breached, flooding the city of New Orleans. More than one thousand people died, the vast majority of whom were African-American, or elderly, or lived in poverty.

The world learned rapidly how the physical and human geography of that city contributed to the disastrous situation. That the coastal city lies below sea level is not unique; so is most of the

20

Netherlands. But hurricanes do not strike Northern Europe, and the low "bowl" shape of New Orleans, combined with the severity of Katrina, exacerbated the flooding. Although the hurricane had been foreseen for days and evacuations were ordered, tens of thousands remained behind, in particular low socioeconomic status (SES) African-Americans and the elderly. They knew the storm was coming, but most survivors repeatedly stated that they had no way to leave and nowhere to go.

Geography mattered in shaping the Katrina disaster, and in the days and weeks that followed, people scoured television and the Internet for maps and imagery of the event. The public saw that spatial data could help them both explain and understand the disaster, and create a cognitive geographical context. We looked at maps that showed how few highways were passable and undamaged. We viewed census data to understand where low-SES African-Americans and the elderly had lived in the city, and we used GIS to overlay other maps that showed the inundated sections.[4] We learned that several of the levees had been built directly over geologic faults, locations that had contributed to their "sinking" over time and resulting in their being lower than originally intended. We became acutely aware of how a hurricane in the Gulf of Mexico could cause higher gasoline prices throughout the United States and the world.

Maps, mapping, GIS, and the spatial thinking of others helped us understand the disaster, by illustrating and explaining the human and physical geography of New Orleans. In the absence of reliable or accurate cognitive maps of the region, we were at a loss to appreciate the situation. Through GIS, we could visualize dozens of distinctive spatial data layers, layers that helped to create the whole story when we saw them together (figure 1).[5]

GIS can make a significant and substantial contribution to building geographic awareness and understanding any place in the world. To learn where New Orleans was located we didn't look at a single, static map. We looked at dozens of dynamic maps, and not only ones that others had already made. Some we could customize, pan and zoom around, turn layers on and off, query and modify their display. We could make our own to share, communicate, and from which learn. New Orleans or the Gulf Coast isn't the local or regional neighborhood for most of us, but these experiences created a long-lasting and profound impression on our awareness of that place.

> Even the specific type of direct interaction can have an effect on both the form and the contents of our cognitive representations of the world. There is a major difference between active and passive ways of exploring a place. The world experienced by someone on foot or on a bicycle is very different from that of a passenger in a bus or car. It is not simply the mode of travel that is important: It is the person's ability to control the spatial sequence and speed of his experience. We develop a very restricted view of a place if we are unable to satisfy the urge to look around the corner. For most people, parts of their world are actively known and parts are known only passively. The passive world is that of the straight and the narrow while the active world is one that we feel at home in" (Downs and Stea 1977, p. 77).

We have only begun to understand the aftermath of Hurricane Katrina. Now we question the integrity of the levees themselves, the long-term impacts of the evacuee diaspora across the United

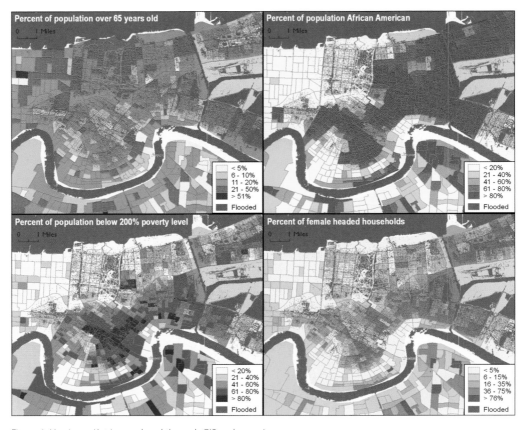

Figure 1. Hurricane Katrina explored through GIS and mapping.
Data courtesy of Center for International Earth Science Information Network (CIESIN), Socioeconomic Data and Applications Center (SEDAC), and Federal Emergency Management Agency (FEMA).

States, and how New Orleans and other affected areas will choose to rebuild their communities. Throughout these analyses, geography will continue to inform and affect the research and decisions. Undeniably GIS and mapping have a role in disaster management itself, but they also provide insights and answers for teaching and learning as well.

Critical, multidimensional thinking and problem solving

College years are a time to build, refine, and apply skills of problem solving and critical thinking. One of the shorter and more succinct definitions of critical thinking is "deciding rationally what or what not to believe" (Norris 1985). Thinking critically is a reasoning process that incorporates stages of categorization, interpretation, analysis, evaluation, assessment, inference, explanation, and other skills applied rationally (Faccione 2004; Fisher 2001). Virtually every institute of higher education strives to have its students learn, practice, and apply these skills.

There is a growing awareness of the value of spatial or geographic thinking to multidimensional critical thinking and problem solving. Inquiry-based education is a defining paradigm of college-level instruction in the United States today. Problem solving is an important part of inquiry-based learning: the "problems" we ask students to grapple with do not have single, numerical answers, or even single correct ones. Instead, we ask students to delve deeper into the reasons why things are the way they are and to tease out complex interrelated elements of connected systems. History students explore why the Roman Empire stopped where it did. Biologists explain why reconstructed wetlands vary in their regrowth patterns. Religion majors investigate the tensions between different groups sharing the same sacred space. Political science students consider the dilemma of redistricting a state's voting districts.

In every one of these situations, the "solution" involves a spatial or geographic understanding of why things are the way they are, how they got to be that way, and what would have to be different to change the results. Students are involved in a critical thinking process, in which they learn to understand the aspects of an issue so thoroughly that they reach a higher-order, synthesis level of knowledge (Bloom 1984). They appreciate the complex nature of the problem, know how the components are interrelated, and recognize how manipulation of one element will affect the others.

> After all, in order to solve a problem one must be able to alter the structure that the situation spontaneously presents to the mind. To perceive is to grasp the salient features of a given state of affairs; but to solve a problem is to find, in that state of affairs, ways of altering relations accents, groupings, selections, and so on in such a way that the new pattern yields the desired solution (Arnheim 1969, p. 194).

Problem solving involves arranging and modifying the variables until a satisfactory solution, or understanding, emerges. Maps, mapping, and GIS facilitate and support the steps of multidimensional, critical, and spatial thinking and problem solving. In the next section we discuss this in specific ways.

Categorizing and interpreting

We begin our understanding of a problem by organizing or categorizing its elements in our head. In many situations, those elements themselves may be abstract ones with which we have no direct or experiential connection (the Roman Empire, for example). Whether cognitively within one's head, with pen and paper, or with digital technologies, mapping forces us to abstract, discriminate, generalize, and simplify reality.

When we choose to produce or generate a GIS map, we decide how to represent that reality in a two-dimensional perspective. Such decisions (i.e., how to show a feature on a map; determining whether it is discrete or continuous, categorical or numerical, aggregated or singular, absolute or fuzzy, and so on) shape and impact our formative understanding of that "ground truth" and is a learning process in itself. Deciding exactly how something is to be shown on a map forces one to think about the nature of that phenomenon. Can a hurricane's path be depicted as a single line?

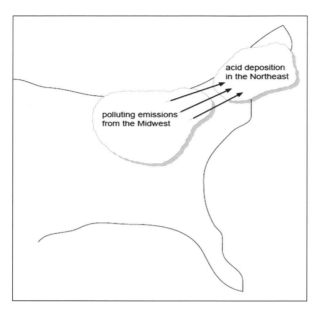

Figure 2. An example of a cognitive map showing approximate areas of midwestern air pollution and northeastern acidic deposition.

How accurate could it ever be to represent the Roman Empire as an area on a map with definite boundaries? Could one have jumped in and out of the Empire by straddling that line? The act of drawing, or representing phenomena as points, lines, or areas, reveals one's understanding of the phenomenon itself, whether it is a road network, a topographic profile of a mountain range, air pollution around a city, or a social kinship structure.

Furthermore, once we see absolute patterns of data in a map, we automatically adjust our previous relative, cognitive maps of that information. For example, students may have an image in their heads that "air pollution in the Midwest leads to acidic deposition (or acid rain) in the Northeast." When we create an absolute map of this data, such as the locations of sulfate depositions across the continental United States, we calibrate our relative and vague mental map that the original statement generates (figure 2) to one that is more accurate (figure 3).[6] With these images in hand, we can begin to decide how valid our initial statement was or was not.

GIS enables us to create countless combinations of data layers, from any number of disparate sources. When the software recognizes where each data layer is actually situated on the surface of the earth, it can align them all and stack them, spatially coincident, on top of one another. Viewing a stacked set of data layers may seem disorienting at first, as our brains compare the new reality with our previous cognitive map (e.g., "I always thought x and y were a lot closer to each other"), but the visualization may support new discoveries and insights or prompt fresh questions (e.g., "That helps explain why x and y aren't as similar as I thought they should be").

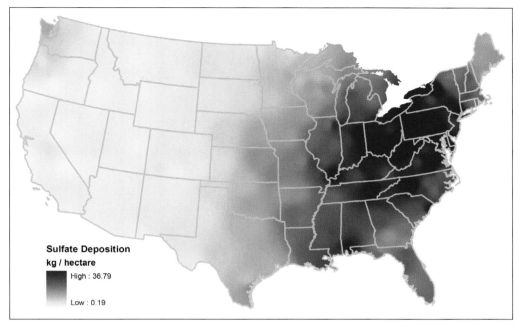

Figure 3. Sulfate deposition amounts, measured in kilograms per hectare, in the contiguous United States. Data is collected and distributed by the National Atmospheric Deposition Program/National Trends Network. The wet deposition at each site is calculated by multiplying the precipitation-weighted mean concentrations by the total annual precipitation at the site, then spatially interpolated using the inverse distance weighted (IDW) algorithm. This data is available at the NADP Web site (nadp.sws.uiuc.edu).

Data courtesy of National Atmospheric Deposition Program (NRSP-3), 2006; and ESRI Data & Maps 2004.

The act of combining previously distinct data layers, whether through a simple and direct visual overlay or a more sophisticated locationally based analysis, is one of the key features of GIS that distinguishes it from traditional mapping (Sui 1995). Because we are able to manipulate data layers, individually or with respect to any combination of others, we can generate maps to display phenomena or spatial relationships that would not otherwise be obvious. Though the different pieces of information may have existed as stand-alone knowledge in our heads, by visually combining the layers we can appreciate the influence and effects that one may have on another.

One example of this is a simple image of the political and cultural variability over the continent of Africa (figure 4). By seeing the ways in which political boundaries cut across areas with profoundly rich ethnic variability, we begin to understand current civil and international conflicts and to put past ones in their context. With the addition of other "layers" of data (figures 5–7), we begin to appreciate how the landforms and environments of Africa may have influenced its cultures, governance, development, and so on. If the viewer is thinking critically and spatially, the visualization can prompt further questions. Do climate and land cover, or climate and land use, correlate with cultural variability? How might the various transportation networks (trains, rivers, roads) correspond with these patterns? Are ethnic patterns coincident with language or religion patterns?

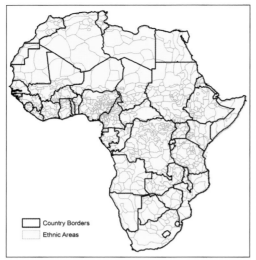

Figure 4. African country boundaries overlaid on distinctive ethnic areas. These areas were originally drawn by George Peter Murdock in his book *Africa: Its Peoples and their Cultural History* (1959, McGraw Hill).

Redrawn from George Peter Murdock, Africa: Its Peoples and Their Cultural History (McGraw Hill 1959). Additional data courtesy of ESRI Data & Maps 2004.

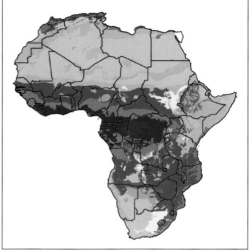

Figure 5. African country boundaries overlaid on vegetation.

Data courtesy of GEO Data Portal, United Nations Environment Programme 2004, taken from Frank White, "Vegetation of Africa," Natural Resources Research Report XX (1983); U.N. Educational, Scientific and Cultural Organization. Additional data courtesy of ESRI Data & Maps 2004.

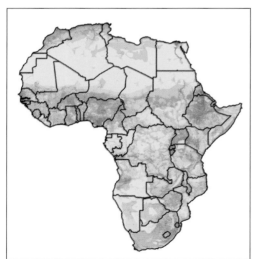

Figure 6. African country boundaries overlaid on estimated population density in the year 2000. Darker colors indicate a greater population density.

From "Gridded Population of the World, Version 2." Courtesy of Center for International Earth Science Information Network. World Resources Institute, 2000.

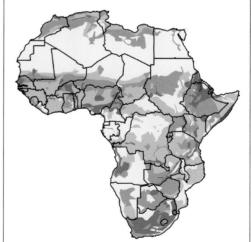

Figure 7. African country boundaries overlaid on soil areas ranked for human-caused levels of degradation. Darker colors indicate soil that is more highly degraded.

From The Global Assessment of Human Induced Soil Degradation Digital Database. Courtesy of UNEP/GRID-Geneva. Additional data courtesy of ESRI Data & Maps 2004.

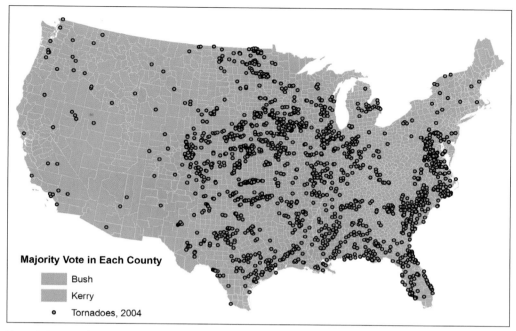

Figure 8. An example of mapping two unrelated data variables, with no known correlation. County-level results of the 2004 presidential election, and the location of tornadoes in the year 2004.

Data courtesy of Dave Leip (www.uselectionatlas.org), NOAA, the National Atlas, and ESRI Data & Maps.

Has the distribution of various ethnic groups changed over the last fifty years, and if so, where and how?

Though the mapping software permits us to line up or stack any number of layers (as long as we know where the data occurs on the earth's surface) there may, in fact, be no relationship or correlation at all between the data layers (figure 8). GIS' ability to enable simultaneous visualization of data from widely different and wholly unrelated data sources can be both a problem and an opportunity if used to teach critical visual skills. Instead of assuming a map is accurate, which noncritical consumers are unfortunately apt to do, spatially literate students learn to question the map, the data it represents, and the relationships it suggests.

Analyzing and evaluating

Our ability to interact with and manipulate GIS-based maps of the Hurricane Katrina region increased our awareness of the region's geography. When students assume control over the scale, display, and organization of the data, they are empowered to gain deeper knowledge. The potential for students to learn varies quantitatively and qualitatively along a continuum, anchored at one end by educational experiences such as viewing static maps, and at the other by working hands-on with

GIS software to create maps. The world of Internet-based maps has recognized this: the popularity of the technology to customize Google Maps has been remarkable.[7]

> Interaction is fundamental to spatial visualization. The process by which a user explores, correlates, and comprehends spatial data is by its nature interactive and iterative. Users benefit from the ability to modify interactively the parameters of their problem and to observe the effects in real time (Medyckyj-Scott 1994, p. 210).

Knowing how and why we would want to manipulate spatial data layers requires us to think spatially. Spatial thinking has been defined as the knowledge, skills, and habits of mind to use concepts of space, tools of representation (such as maps), and processes of reasoning to structure problems, find answers, and express solutions to these problems (National Academies Press 2005). We focus here on one aspect of spatial thinking termed spatial relations, a skill set that includes recognizing, associating, and correlating spatial patterns and distributions.

In their landmark 2005 report on the potential of GIS to support spatial thinking, the National Academy notes that users of GIS and spatial data are applying important skills that make up a "cognitive spatial toolkit" (National Academies Press 2005; Golledge and Stimson 1997). In fact, cognitive maps, whose generation and recall is affected by these spatial relations, can be construed as an "internal GIS" (Golledge and Stimson 1997). In both cognitive and GIS-based maps, information is symbolized and coded, categorized and organized, manipulated and represented in a variety of manners. However, when we mentally overlay two or more layers of information in our heads, the results tend to be "error prone"—unless people are allowed to actually see pictures of those layers after they are combined. Then they are better able to recall information from the separate, individual layers (Golledge 1995). As Golledge reports, when we use GIS to replicate some of our mental processes of visually correlating and associating layers of information, the process is not subject to the variability that our cognitive manipulation introduces, and GIS will execute the tasks more "quickly and accurately."

For example, proximity and connectivity are two different functions that explain spatial patterns or relationships between phenomena. These concepts are not likely to emerge as specific topics in a class, but they can be used to support or explain stories from any number of courses. Imagine a typical student in a nineteenth-century American history course examining patterns of slavery, with prior knowledge simply that the Southern states had high proportions of enslaved African-Americans. We wish, however, for the student to refine her knowledge. Was slavery distributed uniformly throughout the South? If not, what factors influenced where and to what degree slavery occurred, and why?

Maps and GIS can extend the student's prior knowledge, filling in geographically important pieces of the story. By considering data layers such as cotton production, soil quality, farm size, transportation routes, and the distribution of enslaved African-Americans, a student could deduce that slave populations were spatially concentrated in and around townships and counties with particularly high cotton production, which in turn lay in the areas of the South with the most fertile soil. These areas had the largest plantations and were connected by extensive transportation networks to coastal cities to ship the cotton overseas. Here, the spatial concepts of proximity and connectivity are fundamental to understanding the distribution of slavery, and ones that GIS illustrates effectively. Using these mapped layers all stacked together, students have a visual entry into the complex geography of slavery and pre–Civil War America, and may understand for themselves how maps and geography can support or refute historical stories.[8]

Spatial relations skills may represent the one area in which GIS is most useful in a liberal arts context, illustrated by intuitive connections with critical thinking skills. In a series of steps that follow a generalized process of inquiry, we take data, visualize it, and examine the patterns, if any, presented. We may also manipulate the data to consider it from different perspectives, and test whether the patterns of correlation that the visual image suggests are indeed valid or significant ones. We can generate visible hypotheses, with multiple dimensions, and systematically assess them to decide whether the spatial patterns or relationships we see really exist (Arnheim 1969).

Making inferences

In the learning process, students move from observing and analyzing to creating new knowledge, exemplified by Bloom's Taxonomy of Learning, and we can envision a parallel track of increasingly sophisticated spatial thinking and GIS use (figure 9) (Bloom and Krathwohl 1956).

GIS accelerates spatial thinking and movement along the path to the synthesis of knowledge because it supports our understanding of large and multidimensional data, obtained from different sources, and allows us to process and visualize this data quickly and iteratively. In this way, GIS establishes the conditions that support higher-order reasoning and support our making inferences and conjecturing alternatives. This is especially important when we don't have the luxury of experiencing a place directly or visually. For example, we are now more cognizant of both the physical

Table 1. *The relationship between ArcView GIS, Bloom's taxonomy, and the geographical route of inquiry (after West, 1998).*

	Level of Complexity	*ArcView* Functions	Bloom's Taxonomy	Geographical Inquiry	
Data manipulation	Generating new information	Layout	Evaluation	How Ought?	Reasoning with geography
Data manipulation	Spatial analysis	Query Builder/ Select by Theme	Synthesis	What Impact?	Reasoning with geography; Reasoning in geography
Data management	Data analysis	Chart/View/Query Builder/Select By Theme	Analysis	How/Why?	Reasoning with geography; Reasoning in geography
Data management			Application		Reasoning with geography; Reasoning in geography
Data management; Data encoding	Data	Theme/View/Table of Attributes	Comprehension	Where?	Reasoning in geography
Data management; Data encoding	Datum	Attribute	Knowledge	What?	Reasoning in geography

Figure 9. Connecting GIS with Bloom's Taxonomy and a geographical route of learning.

Reprinted from Bryan West, "Student Thinking and the Impact of GIS on Thinking Skills and Motivation," *Journal of Geography*, November/December 2003. Courtesy of the National Council for Geographic Education.

and human geography patterns of New Orleans. We can now envision what might happen if a storm like Hurricane Katrina struck Miami or Havana or Galveston, and anticipate what would be similar and what would be different. If students experience a successful spatially based inquiry in one course or with one incident, they may learn quickly to make informed inferences beyond their sphere of direct experiences.

Furthermore, through GIS-supported spatial thinking, we can readily appreciate the interconnectedness among disciplines, and also between the conceptual realm of our classrooms and the practical realm of our world, something too often tacit on a college campus. GIS and spatial thinking (as well as geography) can function as a common denominator between subjects, making explicit the true interdisciplinary nature of the world. This too distinguishes the best integrated learning styles from the mediocre (Huber and Hutchings 2004).

Concluding thoughts

Promoting spatial thinking, developing global awareness, and supporting the critical thinking process represent several of our objectives for integrating GIS into liberal arts curricula. Across all areas of the curriculum, we want our students to absorb, digest, evaluate, interpret, and synthesize information. These skills are at the foundation of a "liberal arts" education because successful and effective critical thinking "liberates" a student from the limitations of always believing what they are told, what they read, what they hear, and what they see (Faccione 2004). Such thinking empowers a student to question authority in a rational and systematic approach. Connecting spatial learning to these goals, through maps, mapping, and GIS, accelerates our progress to achieving those objectives.

While it is evident that GIS has the potential to fill a role in the learning process, the path is neither straightforward nor inevitable for numerous reasons. All of the major software packages are only now beginning to incorporate time-based representations and analyses. Universal access to spatial data may eventually benefit all types of learners (Golledge, Rice, and Jacobson 2005). We lack satisfying methods to incorporate fuzzy datasets and visualize error and uncertainty. Only at the cutting edge of cartographic research can we explore spatial data representation in innovative statistical space.[9]

However, by far the greatest barriers and impediments to using GIS as a pioneering pedagogical tool reside in the ability of faculty, at all educational levels, to be aware of, appreciate, understand, use, and teach GIS and mapping themselves. While we proceed to make small steps toward integrating GIS in the classroom, we lag far behind in our teaching of spatial thinking (National Academies Press 2005). Acquisition of software skills does not necessarily equate with abilities to identify, pose, or answer sophisticated spatially based questions of the data. Educators require explicit training and guidance in spatial thinking in order to take advantage of geospatial technologies in their classrooms (Baker and White 2003; Alibrandi 2003; Kerski 2003). Preparing and expecting nongeography faculty to ask geographic questions of their datasets (or to collect spatially based data) is a grand task in itself (Audet and Paris 1997). When college faculty themselves are not accustomed to thinking spatially, they are unlikely to model the behavior for their students, to elicit spatially based inquiries, and to assess spatially based responses. Though we believe GIS establishes the conditions that support higher-order reasoning, there is no guarantee that faculty or students will necessarily make that leap through their use of the software alone. Fortunately, the importance of spatial thinking is increasingly recognized for younger students, and one day there may indeed be spatial-thinking-based standards embedded into other educational goals (National Academies Press 2005). Simultaneously, the availability and accessibility of Internet-based mapping applications is rapidly breaking down the perception that "industrial strength" software is the only option.

Meanwhile,

> Geography is a fertile ground for crossing the traditional boundaries of science, social theory, technology, and the humanities, and capacious imaginations will be required to realize the potential of GIS to better understand—and in some limited cases, even solve—scientific problems, illuminate social injustices, and feed the human spirit (Sui 2004).

Faculty and students in higher education have those capacious imaginations. They live and learn in an environment that welcomes intellectual curiosity and demands the ability to think, write, read, and speak about an increasingly interdisciplinary world. It is our hope that educators and students recognize the utility of GIS, learn to use it wisely, and run with it.

Notes

1. See, for example, news.nationalgeographic.com/news/2006/05/0502_060502_geography.html.
2. For the story, see www.cnn.com/2005/US/09/06/katrina.charleston
3. Google Maps, Google Earth, Windows Live Local, and NASA's World Wind are a few examples of browser-based mapping services and virtual globes.
4. Maps from *The New York Times*: www.nytimes.com/packages/html/national/2005_HURRICANEKATRINA_GRAPHIC/index.html.
5. See, for example, this Web site from the New York Times Katrina Maps: www.nytimes.com/packages/html/national/2005_HURRICANEKATRINA_GRAPHIC/index.html. NOAA's Impact Assessment Maps (www.ncddc.noaa.gov/Katrina-2005/InteractiveMaps) and a Hurricane Katrina Disaster viewer (apps.arcwebservices.com/sc/hurricane_viewer/index.html)
6. The National Atmospheric Deposition Program maintains maps of both concentration and deposition of sulfate, nitrate, and ammonium concentrations. GIS-ready data can be downloaded as well at nadp.sws.uiuc.edu.
7. See googlemapsmania.blogspot.com and www.google.com/apis/maps, and www.googleearthhacks.com.
8. See Molefi Asante (1991) and Larry Gara (1996).
9. See, for example, the work of geographers Andre Skupin and Sara Fabrikant.

References

Alibrandi, Marsha. 2003. *GIS in the classroom: Using geographic information systems in social studies and environmental studies.* Portsmouth, N.H.: Heinemann Publishers.

Arnheim, Rudolf. 1969. *Visual thinking.* Berkeley, Calif.: University of California Press.

Asante, Molefi. 1991. *The historical and cultural atlas of African Americans.* New York: Macmillan.

Audet, R. H. and J. Paris. 1997. GIS implementation model for schools: Assessing the critical concerns. *Journal of Geography* 96(6): 293–300.

Baker, T. R., and S. H. White. 2003. The effects of GIS on students' attitudes, self-efficacy, and achievement in middle school science classrooms. *Journal of Geography* 102:243–54.

Bloom, Benjamin. 1984. *Taxonomy of educational objectives.* Boston, Mass.: Allyn and Bacon Publishers.

Bloom, Benjamin S., and David R. Krathwohl. 1956. *Taxonomy of educational objectives: The classification of educational goals, by a committee of college and university examiners. Handbook I: Cognitive domain.* New York: Longmans, Green.

de Blij, Harm. 2005. *Why geography matters: Three challenges facing America: Climate change, the rise of China, and global terrorism.* New York: Oxford University Press.

Downs, Roger, and David Stea. 1977. *Maps in minds, reflections on cognitive mapping.* New York: Harper and Row.

Faccione, Peter A. 2004. *Critical thinking: What it is and why it counts. Insight assessment.* Millbrae, Calif.; California Academic Press.

Fisher, Alec. 2001. *Critical thinking, an introduction.* New York: Cambridge University Press.

Gara, Larry. 1996. *The Liberty Line: The legend of the Underground Railroad.* Lexington, Ky.: University Press of Kentucky.

Golledge, Reginald, M. Rice, and R. D. Jacobson. 2005. A commentary on the use of touch for accessing on-screen spatial representations: The process of experiencing haptic maps and graphics. *Professional Geographer* 57:339–49.

Golledge, Reginald, and Robert J. Stimson. 1997. *Spatial behavior: A geographic perspective.* New York: Guilford Press.

Golledge, R. G. 1995. Reasoning and inference in spatial knowledge acquisition: The cognitive map as an internalized GIS (Final Report No. NSF Grant SES-92-07836): Department of Geography, Research Unit on Spatial Cognition and Choice, University of California, Santa Barbara.

Huber, Mary Taylor, and Pat Hutchings. 2004. *Integrating learning: Mapping the terrain.* Washington, D.C.: Association of American Colleges and Universities.

Kerski, Joseph J. 2003. The implementation and effectiveness of geographic information systems technology in secondary education. *Journal of Geography* 102(3): 128–37.

Lloyd, Robert. 1997. *Spatial cognition, geographic environments.* Norwell, Mass.: Kluwer Academic Publishers.

Medyckyj-Scott, D. 1994. Visualization and human-computer interaction in GIS. In *Visualization in geographical information systems,* ed. Hearnshaw and Unwin, 200–11. New York: John Wiley & Sons.

National Academies Press. 2005. *Learning to think spatially: GIS as a support system in the K-12 curriculum.* Washington, D.C.: National Research Council.

Norris, Stephen P. 1985. Synthesis of research on critical thinking. *Educational Leadership* 42(8): 40–45.

Porter, Burton. 2002. *The voice of reason: Fundamentals of critical thinking.* New York: Oxford University Press.

Sui, D. Z. 1995. A pedagogic framework to link GIS to the intellectual core of geography. *Journal of Geography* 94:578–91.

———. 2004. GIS, cartography, and the "Third Culture": Geographic imaginations in the computer age. *Professional Geographer* 56(1): 62–72: 67.

About the authors

Diana Stuart Sinton is the chief program officer for Latitude, a collaborative GIS initiative at NITLE, the National Institute for Technology and Liberal Education (www.nitle.org). NITLE works with its participating colleges to promote the effective use of technology and catalyze innovative teaching. NITLE is an initiative of Ithaka (www.ithaka.org) and receives generous funding from The Andrew W. Mellon Foundation. Sinton holds BA, MS, and PhD degrees from Middlebury College and Oregon State University. She has taught landscape ecology, geography, and GIS at the University of Rhode Island and Alfred University, and has conducted research in the Pacific Northwest and Argentina. Currently she teaches, presents, and writes about the role of GIS in higher education, and takes time to make maps on occasion.

Sarah Witham Bednarz is Associate Professor of Geography at Texas A&M University, where she teaches courses in human geography and geography education. She directs an NSFG K–12 program called Advancing Geospatial Skills in Science and Social Science, which links undergraduate and graduate students with K–12 schools to promote spatial thinking, GIS, and related technologies.

Chapter 3

Finding narratives of time and space

David J. Staley

Recently, I asked a contemporary society class to "draw globalization" (figure 1). I gave each student a blank map and then asked them to fill it with symbols, lines, words, shapes, colors, and whatever other marks they wanted to include. I have used this exercise many times when teaching introductory history classes; the purpose is to get students to see historical geography as more than just the memorization of places on a map (in much the same way I want them to understand history as more than the memorization of names and dates). I want them to understand geography as the study of the location of events in time and processes in space. More broadly, I want students to grasp the meanings which fill the spaces they are depicting. The best way I know to accomplish this is to have students compose their own maps.

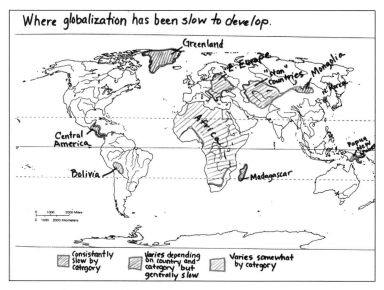

Figure 1. Student's map, "Globalization." Courtesy of Lindsay Kagy.

This is not an extra-credit assignment: when calculating the final semester grade, I give the map the same weight as a written essay. Indeed, I present the assignment as an alternative to writing, as a way for them to exercise other cognitive abilities, other multiple intelligences. In form, this assignment is little different than what I ask students to do when they write an essay. For a written essay, I would ask a question like "What is the impact of globalization?" and expect students to design an argument, with an introduction that announces a thesis, well-formed paragraphs that defend that thesis, and a conclusion. When I ask students to write, I do not want to see a disorganized list of facts but, rather, I want organization and coherence, the facts arranged into a meaningful pattern. Similarly with my map assignment, I want students to arrange and organize visual information into a meaningful pattern.

Good expository writing is often built from narrative: a plot, an organizing principle, a setting, a meaningful arrangement of sentences, propositions, and paragraphs building toward a larger goal or purpose. The map exercise asks for nothing less; I evaluate the students based on criteria similar to a written essay. Their maps need to be clear and accurate, compositionally sound, and full of information. Most importantly, their maps must be organized and coherent. They cannot be a haphazard mélange of symbols and geographic facts without any connection to one another. Their maps must be organized around a "story." Indeed, their maps must be organized around a *spatial narrative.*

In asking students to create their own maps, I invite them to design their own stories, their own spatial narratives. And not just in their roles as producers of maps; when they are looking at maps as well—acting as consumers of geographical information—I want them to be able to discern the story of the map. Geographic information systems offer all of our students the opportunity to create their own visual stories, their own spatial narratives that can be as rigorous and well-formed as written narratives.

A GIS map as a site for narrative

My first experience with a GIS map occurred at a conference in 1997. A team of three graduate students in history presented a GIS display showing voting behavior in Massachusetts in the 1860 election. The team was trying to understand the relationship between distance to major polling areas and voting behavior. Their complex, data-rich map was the centerpiece of their presentation; their analysis was contained in the map, and, in fact, their verbal descriptions served as an illustration of the image, reversing the typical cognitive division of labor between words and images. Afterward, I spoke to the students and informed them that if they were my graduate students, I would accept a—admittedly more complex—version of this map in lieu of a written dissertation. That is, I would allow these students to organize their data not in the form of a textual object but as a dynamic visual object as a way to convey their historical narrative.

Geographic information systems are one example of a suite of technologies—from data mining to immersive virtual reality displays to complex mathematical spaces—that have been collectively labeled "information visualizations." A visualization is any graphic that organizes data into spatial form for purposes of display, analysis, interpretation, and communication. Such cognitive objects are hardly new: maps, scientific illustrations, and statistical graphics all predate the computer by centuries. Nevertheless, since the early 1990s, the number of information visualizations appearing in scientific publications, academic conferences, and in Web pages has exploded (Chen 2002).

Moreover, the function of these images has also undergone significant change. They play a new role in the thinking process. Digital information visualizations, especially, are different from their paper predecessors, and not simply because they are computer generated. A GIS map or a data mining display are not simply attractive visual aids used to illustrate a textual presentation—or, less charitably, a dumbed-down distraction from the "real" information contained in the text. These computerized displays are used by both scientists and mathematicians as tools to explore data. Numerical computations displayed in dynamic visual form—used to study everything from strange attractors in physics to complex mathematical spaces to bioinformatics—can be explored as if they were an uncharted territory. These visual displays are not illustrations of the real scientific work; in many cases, these displays are the real work.

Another important difference between computerized information visualizations and earlier types of graphics is that the former often serve as the reader's (or viewer's) main interface with the information. This is a nontrivial point: the theorist W. J. T. Mitchell has observed that, in Western culture at least, images frequently serve a secondary role in the "division of labor" between words and pictures (Mitchell 1994). Words, text, and prose are the main carriers of information; a visual display is, by definition, an illustration, something that supports the more important ideas contained in the text. That is, the text is the reader's main interface with the data: words, sentences, paragraphs, and prose give form to shapeless information. Information graphics occupy a subordinate position within this textual space. With information visualization, the polarities are reversed: the visual display is the main interface with the data, that which gives shape and form to the information.

The convergence of four trends in the 1980s may account for both the dramatic increase in the number of information visualizations and their change in cognitive status. First was the appearance of fourth-generation computers, and especially the graphics capabilities enabled by these tools. More than simply a word processor or a calculating machine, these tools could display rich data graphics, which changed how scientists, mathematicians, and some artists worked.

This led to the second trend: the emergence of the "new sciences" of complexity theory and fractal geometry. Physicists and mathematicians began to use computers to display numerical data in phase space and other types of mathematical spaces, and thus were able to see meaningful patterns in the data that had escaped awareness when these were presented in tabular form or in a spreadsheet. Perhaps the most famous example was the popular Mandelbrot set, a complex example of fractal geometry. The algorithm that produced the complex shape had long been recognized by mathematicians, but was always dimly understood, until powerful graphics capabilities allowed Mandelbrot to visually represent it (figure 2). Because of the work produced by these scientists and mathematicians, data exploration through computerized visual representation has become a legitimate form of scholarly inquiry.

Third, Edward Tufte published his groundbreaking *The Visual Display of Quantitative Information* (Tufte 1982). Among its many other virtues, the book made a powerful and persuasive case that data graphics needed to be composed with the same attention to clarity, rigor, and thoughtfulness that we attend to written prose. Tufte established a set of principles guiding the composition of graphics—a visual Strunk and White—and in so doing reminded his readers that data graphics are not simply amusing diversions that break up the monotony of long blocks of dry text. Graphics are rigorous ways to display, explore, and communicate rich and meaningful information—a well-formed graphic could contain a narrative.

Fourth, Howard Gardner published *Frames of Mind* (Gardner 1983), the book that first defined the term "multiple intelligences." Rejecting the long-held belief that human intelligence could be measured by a single number (as with an IQ test), Gardner maintained that every human was a collection of several different intelligences, including linguistic, mathematical, spatial, and kinesthetic—each of varying degrees and aptitudes. Some people are especially strong in linguistic skill, others in the kind of intelligence it takes to move their bodies, others at thinking through images. Today, Gardner's ideas have become standard among psychologists and educators; teachers are taught that some students are better able to acquire information visually rather than through traditional forms of instruction, such as oral lectures or written textbooks.

Taken together, the convergence of these four trends created the context that has allowed information visualizations to thrive. More importantly, thinking visually and spatially is increasingly

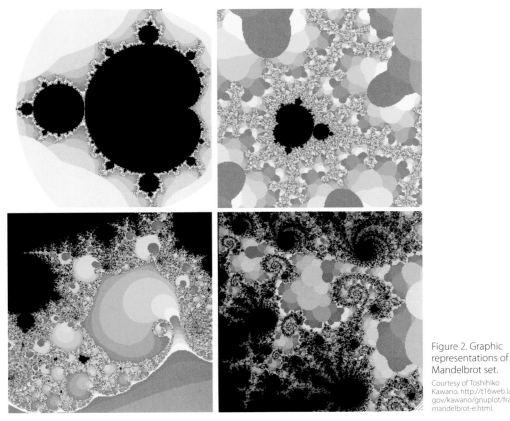

Figure 2. Graphic representations of Mandelbrot set.

Courtesy of Toshihiko Kawano, http://t16web.lanl.gov/kawano/gnuplot/fractal/mandelbrot-e.html.

understood today to be a legitimate form of inquiry and scholarly communication. In this setting, it is much easier to accept the idea that a visual display of information, like a GIS map, can be the main carrier of the meaningful information, an alternative to written prose as a site for narrative.

Spatial narratives

When we use the word narrative, we typically think of a story, the telling of events that unfold in time, usually leading to some sort of climax or conclusion. The term is usually applied to stories we tell or write in words, but we can also describe many maps as serving a narrative function. Indeed, the curators of an exhibit at the Newberry Library noted that "Every Map Tells a Story" (Akerman 2000). The idea that a map can convey narrative might sound counterintuitive to some: we usually think of narrative as a linear plot that unfolds in time, while maps are concerned with the depiction of multidimensional space. This is an important distinction, and I will have more to say on this point below (namely that space itself can be a type of narrative). For now I would point out that maps—and not just GIS maps—have often been used to depict the movement of time.

As the historian Michel de Certeau observed, spatial stories often fall into one of two categories: the tour and the tableau (de Certeau 1984). The tour refers to a depiction of an itinerary through space, as when a map shows direction or movement across a landscape. Consider a common experience: you are giving directions to a friend, and draw a map that traces a path through space that instructs your friend, "first, go by the gas station, then turn right and go a mile to the video store," and so on. Such a map involves drawing a line through space, setting a direction, a temporal path toward a final goal or outcome. When I ask my students to "draw the Black Death," very often they depict the itinerary of the disease, from its origins in the Asian steppe to its spread westward, and its subsequent decades-long swirl around Europe. The students trace lines through space, but they will often affix dates and notable events along these lines, thereby situating the spread of the disease in a temporal as well as a spatial context.

There are numerous examples of such temporal maps from the past. Certeau observes that "the first medieval maps included only the rectilinear marking out of itineraries [pilgrimages, mostly], along with the stops one was to make" (de Certeau 1984, p. 120). Aztec and Toltec cartographic histories were visual records of notable events from the past arranged within a cartographic space (figure 3) (see Boone 1994). Battle maps show the movement of troops and their actions within a defined space. Maps in historical atlases very often convey the passage of time, as when a sequential series of maps depicts change through time and space. In the case of these maps, as with my Black Death example, the temporal dimensions seem to take precedence over the spatial qualities, or at the very least the temporal and the spatial exist in harmony within one map.

In the cartographic histories, GIS lets the mapmaker depict change and movement through time, telling a story chronologically. GIS also lets us show movement through space, depicting tours and itineraries geographically. GIS can also produce iterative maps, a series of maps that reflect different queries of the same database. A sequence of such maps would be able to show change over time, not unlike a series of maps in an atlas. Students clicking on a hyperlink contained within a historical map might see a contemporary map of the same location, the juxtaposition of the two maps

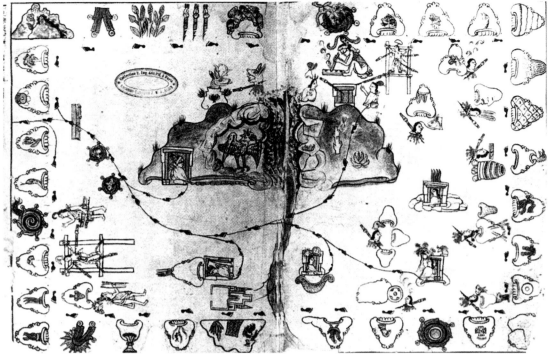

Figure 3. Toltec cartographic history.
Used by permission of Benson Latin American Collection, University of Texas at Austin.

inviting a comparison (see Rumsey and Williams 2002). Animated GIS maps can extend the power of sequentially ordered maps to show change through time. John Brockhaus and Michael Hendricks, for instance, have described animated GIS maps that can depict changes in both battlefield attributes and position (Brockhaus and Hendricks 1998), in effect replacing a series of static maps arrayed in sequence as in an atlas.

The second type of spatial story identified by de Certeau is the tableau, the depiction of a space. Certeau suggests that, between the fifteenth and seventeenth centuries, modern maps were largely emptied of their depictions of itineraries, and instead became vessels for storing knowledge. "Transformed first by Euclidean geometry and then by descriptive geometry, constituted as a formal ensemble of abstract places," argues Certeau,

> [the modern map] is a "theatre" (as one used to call atlases) in which the same system of projection nevertheless juxtaposes two very different elements: the data furnished by a tradition (Ptolemy's Geography, for instance) and those that came from navigators (portulans, for example). . . . The map, a totalizing stage on which elements of diverse origin are brought together to form a tableau of a "state" of geographic knowledge, pushes away into its prehistory or into its posterity . . . the operations of which it is the

result or necessary condition. It remains alone on the stage. The tour describers have disappeared (de Certeau 1984, p. 121).

As a result of this totalization of space, modern maps are less likely to display the temporal narrative common to "tour" maps.

We might define the modern map as a form of synchronic narrative, rather than the diachronic narrative of tour maps. A diachronic narrative describes or depicts change through time, a standard form of history writing; a diachronic map could be said to do the same. A synchronic narrative, on the other hand, describes or depicts the relations within a system at one particular moment in time. The anthropologist Clifford Geertz's method, which has been described as a "thick description" of a culture, is an example of a synchronic narrative. Influenced by both geography and anthropology, the Annales school of history rejected diachronic narratives of the flow of events in favor of a description of long-term systems of spatial and cultural relations. Maps that show systems and relations within a given moment convey narrative and story as much as maps that show change through time.

Therefore, I would disagree with Certeau's characterization of tour maps as the only ones that convey a story. I would argue that the modern synchronic map displays a different type of narrative: a spatial narrative. A map is a two-dimensional object; that is, its syntax is two-dimensional, rather than the one-dimensional syntax of language. Syntax is defined as the rules which link together symbols. In language, syntax refers to the linking together of words in proper order. That order is, regardless of the language, a one-dimensional chain: this word, then this, then this, and so on. The syntax of an image, and I include maps here, is two-dimensional, in that the symbols—words, lines, polygons, colors—link together in a web, not a sequential chain. The geographic information contained in a map "is multidimensional," observes the historical geographer Anne Kelly Knowles, "requiring at least two coordinates to define location in space and at least three if the item being located in space is also being located in time. [Geographic information] is often voluminous, because capturing the geographic extent of a large area and its features can easily generate gigabytes of data. It is often best represented in graphic form, particularly in maps, rather than in tables, charts or text" (Knowles 2002, p. xiv). Since geographic information is itself multidimensional, the story contained within that information must also be multidimensional.

Despite its volume, the information contained in a map takes up a much smaller area than if we were to write out all the information. To see what I mean, take any map and try to translate into language all the information contained in that map. The map with its two- or multidimensional syntax is capable of displaying information simultaneously, rather than in a one-dimensional sequential line as with language. The map can concisely convey the relationship between wholes and parts. If we were to write out all the information—indeed, the narrative of the map—after just a few sentences, the contrast between linear narrative and spatial narrative would become clear.

Thus, in such a two-dimensional space, "narrative" takes a different form than what we would expect in a one-dimensional linguistic story. When considering two- and three-dimensional spaces like a map, we need to broaden our definition of narrative, or at the very least devise a definition for spatial narratives. This step has already been taken by literary theorists interested in the properties of hypertext as a literary form (see Aarseth 1997; Bolter 1991; Landow 1992). "Text on the computer screen" has challenged many of our ideas about linear narrative, ideas that date to Aristotle's formulations about plot and narrative. The emergence of hypertext caused literary theorists to reexamine many of their assumptions about narrative, especially the assumption that narrative need be linear and logocentric. These theorists have observed that hypertext narratives are often organized like maps; these stories exhibit spatial qualities, as they unfold in nonlinear or multilinear ways. Hypertext stories can include visual and aural elements that disrupt our traditional notions of linguistic narrative.

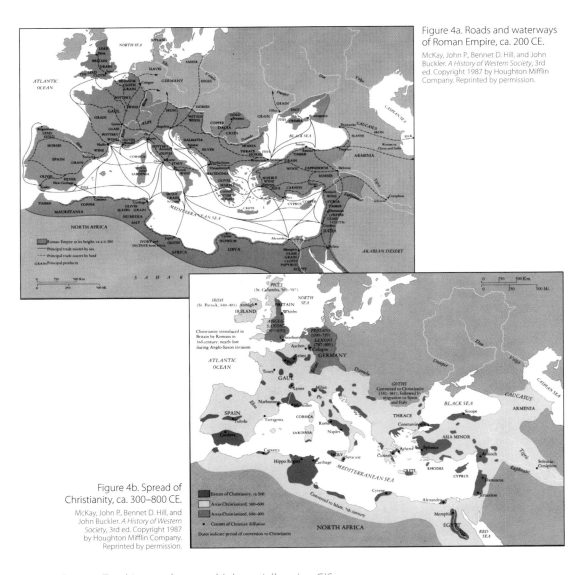

Figure 4a. Roads and waterways of Roman Empire, ca. 200 CE.

McKay, John P., Bennet D. Hill, and John Buckler. *A History of Western Society*, 3rd ed. Copyright 1987 by Houghton Mifflin Company. Reprinted by permission.

Figure 4b. Spread of Christianity, ca. 300–800 CE.

McKay, John P., Bennet D. Hill, and John Buckler. *A History of Western Society*, 3rd ed. Copyright 1987 by Houghton Mifflin Company. Reprinted by permission.

Can we thus continue to call such spatial narratives narratives at all? To my way of thinking, the answer is yes if we understand narrative to mean the organization and ordering of elements into a meaningful pattern. In a written narrative, the elements of the story move along in a linear pattern, with a beginning, middle, and end. In a spatial narrative, as we might find on a map, the pattern can be linear, as we see with tour-style maps, but more often than not there are other forms of meaningful patterns, patterns that unfold in two or more dimensions. In its classic Aristotelian formulation, a central concept in understanding narrative is emplotment. GIS maps represent a kind of multidimensional emplotment: a single story organized from multiple and heterogeneous elements as a spatial totality.[1]

One of my favorite map exercises introduces students to this idea of identifying a meaningful pattern within a spatial narrative. I begin with two map transparencies, one showing the network of roads and waterways that traversed the Roman Empire, the second showing the spread of Christianity from its early concentration in urban areas to its slow diffusion over time into the hinterlands (figure 4). Then, I superimpose one map onto the other and ask the students to tell me what they see. Overlaying the two maps, the relationship between the roads and waterways and the spread of Christianity are made very clear: that Christianity spread along the trade routes that connected cities in the Roman Empire. Students then light upon the historical irony: that one cause of the fall of Rome—the spread of Christianity—was itself facilitated by the very infrastructure of the Roman Empire. I could, of course, attempt to describe this irony to them in words, but until the students see the story unfold via the spatial narrative of the map this historical irony does not register.

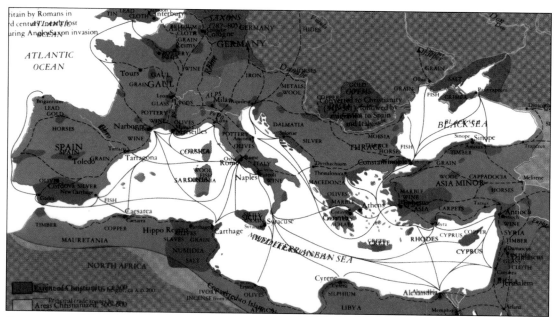

Figure 4c. Overlay of figures 4a and 4b showing relationship between Roman trade routes and spread of Christianity.

McKay, John P., Bennet D. Hill, and John Buckler. A History of Western Society, 3rd ed. Copyright 1987 by Houghton Mifflin Company. Reprinted by permission.

Superimposing one map over another helps us to see relationships between the data in each map that might not be well understood were the maps viewed separately. Geographers refer to this as relational cartography. In the past, this technique of data analysis relied on drawing maps on thin paper, or on transparent film as in the above example. The geographer could then take two maps drawn to the same scale, each depicting a different dataset, and overlay one on top of the other. The resulting visual patterns reveal areas of convergence between the two datasets, suggesting a relationship. Exposing this visual correlation reveals the possibility of a causal relationship, potentially deserving further study (Monmonier 1993). In the example above, students saw trade routes, punctuated by urban hubs, aligned with areas of early Christian influence, suggesting a link between the phenomena.

This example deals with only two maps, with two sets of data. If I were to overlay three or more maps, it would soon be difficult to see through the stack of transparencies. An important feature of GIS technologies—perhaps the most important feature—is their ability to layer a number of maps on top of each other to see relationships between the various datasets. Correlations and combinations of different datasets suggest patterns in those data. For example, the environmental historian Geoff Cunfer has juxtaposed maps locating the strongest dust storms during the Dust Bowl of the 1930s with maps depicting rainfall amounts, the quality of the soil, intensity of farming, and average temperature. The spatial pattern that emerges from this layering of maps reveals that high temperatures and low rainfall—rather than the land-use practices of farmers—created the conditions for the Dust Bowl (Cunfer 2002). Meaningful patterns emerge: the hallmark of what I have been calling the spatial narrative of the map.

Participatory narrative

Computer graphics allow interactivity, meaning the information on the screen can respond to the actions and decisions of the viewer. Interactivity is an especially noteworthy feature of GIS. Unlike a paper map, a viewer can change the data displayed in a GIS map, can alter its perspective, can query different relationships. That is, both the viewer and the creator have control over the story of the map. The interactivity of the computer environment makes GIS maps more open than traditional paper-based maps, where "open" is defined as a customizable cartographic space—the information displayed in the map is as much the result of the viewer's choices as it is the mapmaker's design—in contrast to the "closed" or static space of traditional maps, frozen at publication.

The same literary theorists who explored the impact of the computer on our ideas of narrative in literature often point to the changing relationships between readers and authors and consumers and producers of text. Within an interactive computer environment, as within a self-navigating hypertext link, readers can construct narratives based on their viewing choices. Reading a text in a hypertext environment is like playing a game (like a "Choose your Own Adventure" game or a video game) in that a reader's choices determine what they see next. "The player, then, is a reader, but an unusually powerful reader, for his or her decisions determine what text will appear next," observes the literary theorist Jay David Bolter.

Within each episode, the reader is still compelled to read what the author has written. But the movement between episodes is determined by the responses of the reader, his or her interactions with or intrusions into the text, and the reader's experience of the fiction depends upon these interactions (Bolter 1991, pp. 121–22).

Janet Murray (1997) terms this "participatory narrative," where the reader's choices of materials are an integral part of the narrative; that in effect, the narrative is a co-creation of reader and author. Because the reader is so empowered, no two narratives are exactly the same. Murray was describing virtual reality spaces, but the term participatory narrative is just as applicable to hypertext spaces and, I would argue, to GIS cartographic spaces.

GIS, like hypertext, alters the relationship between map creator and map viewer, allowing the viewer to be a co-creator of the story of the map. In choosing which datasets to draw, which subsets to select, or what magnification to use, viewers of a GIS map have a certain power over what they see. Of course a GIS map requires a cartographer, as a hypertext work requires an author, to assemble the elements. Yet a GIS display is replete with options, and the final map depends on the viewer's choices.

Those choices are not limited to traditional map elements. Viewers navigating a GIS space might also query and click on tables, diagrams, photographs, personal letters, or sound clips, all related to the cartographic space of the map. As Mark Monmonier observes, computerized cartography has moved well beyond simple animation. "Multiple linked windows can provide fuller integration between the cartographic space of the map, the attribute space of scatterplots and time-series graphs, and the virtual reality of two- and three-dimensional photographic images. And expert systems/artificial intelligence frameworks," he predicts, "promise customized presentations, including narratives tailored to a viewer's expertise, interest, or residential history as well as intelligent agents for uncovering meaningful patterns or offering skeptical re-interpretations" (Monmonier n.d.). It would appear that the cartographic space of GIS is not only multidimensional but multimedia as well.

In many ways, a GIS map appears as a cartographic type of ergodic literature. This term *ergodic* was coined by the literary theorist Espen J. Aarseth to mean "a work that requires labor from the reader (or viewer) to create a path" (Montfort 2001). Janet Murray would describe the creator of this material as the procedural author (Murray 1997). The procedural author of a GIS space serves as a kind of "narrative architect" [in Henry Jenkins' useful phrase (Jenkins 2004)], and the viewer serves as the builder of the spatial story. Narrative emerges from the complex relationship between an author's constructions and a viewer's choices. Far from the classic Aristotelian linear narrative, a GIS is a multidimensional narrative space of images and words, of procedural author and active viewer, of story and interactivity. GIS maps are a rich expression of this ergodic literature.

Concluding thoughts

All maps contain stories, whether diachronic or synchronic, temporal or spatial. Students need not employ GIS technologies in order to discover these stories, or indeed to create their own spatial narratives. As Anne Knowles has convincingly demonstrated, relatively simple tools such as paper and pencil, Mylar transparencies, and colored pens are all that are needed to teach students cartographic skills and geographical visualization (Knowles 2000). Like the mapping exercise I described at the beginning of this essay, students can use paper and pencil to compose their own spatial narratives.

GIS technologies make many of these tasks easier, however, and can afford rich narrative opportunities that paper maps cannot offer. With GIS, students manipulate the myriad elements of a map to reveal stories by linking to multimedia objects. As GIS software becomes more user friendly, it will become even easier for nongeography majors to apply spatial thinking and geographic visualization to a host of questions in the humanities, social sciences, and sciences. The increasing use of GIS reminds us that students need not be merely consumers of geographical information but can be active creators, utilizing the openness and user control of the technology to find and share spatial stories and multidimensional insights.

Notes

1.   This is a reformulation of Paul Ricoeur's notion of linear emplotment (see Wood 1991, pp. 21–22).

References

Aarseth, Espen J. 1997. *Cybertext: Perspectives on ergodic literature.* Baltimore: The Johns Hopkins University Press.

Akerman, James. 2000. Every map tells a story. The Thirteenth Kenneth Nebenzahl, Jr., Lectures in the History of Cartography. Mapline 90 (Spring 2000). www.newberry.org/smith/maplinefeature.html.

Bolter, Jay David. 1991. *Writing space: The computer, hypertext, and the history of writing.* Hillsdale, N.J.: L. Erlbaum Associates.

Boone, Elizabeth Hill. 1994. Aztec pictorial histories: Records without words. In *Writing without words: Alternative literacies in Mesoamerica and the Andes,* ed. Elizabeth Hill Boone and Walter D. Mignolo, 50–76. Durham, N.C.: Duke University Press.

Brockhaus, John, and Michael Hendricks. 1998. Abstract: Interactive interpretive battlefield maps. ESRI User Conference. gis.esri.com/library/userconf/proc98/PROCEED/ABSTRACT/A741.HTM.

Chen, Chaomei. 2002. Editorial: Information visualization. *Information visualization 1* (March 2002): 1–4.

Cunfer, Geoff. 2002. Causes of the Dust Bowl. In *Past time, past place: GIS for history,* ed. Anne Kelly Knowles, 93–103. Redlands, Calif.: ESRI Press.

de Certeau, Michel. 1984. Spatial stories. In *The practice of everyday life.* Berkeley, Calif.: University of California Press.

Gardner, Howard. 1983. *Frames of mind: The theory of multiple intelligences.* New York: Basic Books.

Jenkins, Henry. 2004. Game design as narrative architecture. In *First person: New media as story, performance, game,* ed. Noah Wardrip-Fruin and Pat Harrigan. Cambridge: MIT Press.

Knowles, Anne Kelly. 2000. A case for teaching geographic visualization without GIS. *Cartographic Perspective* 36 (Spring 2000): 24–37.

———. 2002. *Past time, past place: GIS for history.* Redlands, Calif.: ESRI Press.

Landow, George P. 1992. *Hypertext: The convergence of contemporary critical theory and technology.* Baltimore: The Johns Hopkins University Press.

Mitchell, W. J. T. 1994. The photographic essay: Four case studies. In *Picture theory.* Chicago: The University of Chicago Press.

Monmonier, Mark. n.d. Abstract: Cartographic narratives, openness, and the new technology. The Thirteenth Kenneth Nebenzahl, Jr., Lectures in the History of Cartography. www.newberry.org/smith/nebenzahl/monmonierbib.html.

———. 1993. *Mapping it out: Expository cartography for the humanities and social sciences.* Chicago: University of Chicago Press.

Montfort, Nick. Cybertext killed the hypertext star. ERB January 2001. www.altx.com/ebr/ebr11/11mon.

Murray, Janet H. 1997. *Hamlet on the holodeck: The future of narrative in cyberspace.* Cambridge: MIT Press.

Rumsey, David, and Meredith Williams. 2002. Historical maps in GIS. In *Past time, past place: GIS for history,* ed. Anne Kelly Knowles, 4–5. Redlands, Calif.: ESRI Press.

Tufte, Edward R. 1982. *The visual display of quantitative information.* Cheshire, Conn.: Graphics Press.

Wood, David, ed. 1991. *On Paul Ricoeur: Narrative and interpretation.* London: Routledge.

About the author

David J. Staley is director of the Goldberg Program for Excellence in Teaching in the department of history at The Ohio State University and is executive director of the American Association for History and Computing. He is the author of *Computers, Visualization and History: How New Technology Will Transform Our Understanding of the Past* (M. E. Sharpe 2003).

groups to establish a master park plan. Their initial goals included defining the function of each city park and suggesting future and specialized park development. Grinnell faculty, staff, and students will engage with the community on park-related issues and assist in the use of GIS, particularly the development of a GIS database of park-related data. Progress has been made on the project—students and community volunteers started by mapping the condition of trees in local parks (figure 4).

Macalester College: Land-use plans in St. Paul, Minnesota. Together, the Geography Department and the Community Service Office at Macalester College have engaged for many years in community collaborations involving GIS. One current project will assist community members in the seventeen planning districts with city-required land-use plans. The Macalester group will help each district to define land-use goals unique to their area and develop maps using GIS as part of this process. Students in a Macalester GIS course will participate in this collaborative project, working with different districts and community members.

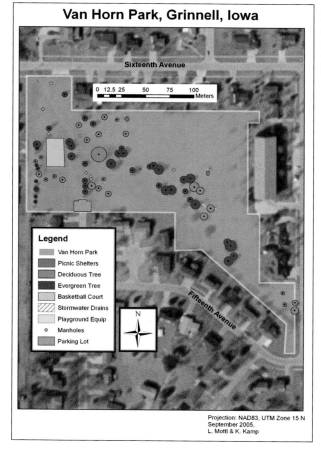

Figure 4. Maps of trees and conditions from the Grinnell College Collaborative Project. ISU GIS Facility and the Iowa Geographic Image Map Server.

Courtesy of ISU GIS Facility and the Iowa Geographic Image Map Server; Larissa Mottl and Kathryn Kamp © Grinnell College 2005.

Middlebury College: Using GIS with local social services. The Middlebury project continues a collaboration between the Champlain Valley Head Start office, the Addison County United Way agency, Middlebury College's Alliance for Civic Engagement office, and the Middlebury College Geography Department. The program's overall objectives are to provide information about available social services, such as health-care and day-care facility locations, through an Internet-based mapping service. Personnel at these local community service agencies are integrally involved with all aspects of this mapping project, from designing the interface to identifying the specific information that the maps will provide—that which they have determined to be most useful and necessary.

Figure 5. Patriot grave at Old Colony Burying Ground. Patti R. Albaugh 2005.

Otterbein College: Mapping Ohio's revolutionary patriots. The Otterbein project seeks to reveal how GIS and local history can assist in meeting geography and history educational requirements (social studies and technology) for middle school students. The project entails collaboration between Otterbein faculty, educational technology students, regional public school teachers, middle school students, local genealogical groups, and the Daughters of the American Revolution. After the conference at Ohio Wesleyan University, a grant was submitted for funds to support the mapping of local Revolutionary War participant graves by Otterbein students and eighth graders (figure 5). The service-learning course was successfully implemented and a Web site was developed in which grave locations are linked to information about the soldiers.[3] The project has helped connect eighth-grade educational benchmarks, including local history, American history, and geography.

Swarthmore College: GIS in the curriculum and the community. The Swarthmore group, consisting of faculty and students, is developing two collaborative GIS projects: a local history project for an economically depressed community in Chester, Pennsylvania, and a watershed pollution and restoration project in suburban Philadelphia. The goals of both projects are to bring GIS into the Swarthmore curriculum, generate funding and support for a campus GIS lab, collaborate with community groups in the region, and address the specific needs of each of the projects.

GIS in service learning

GIS community-based work in higher education has been informed and enhanced by the innovative pedagogy of service-learning. Service-learning is a structured learning process grounded in explicit learning objectives, preparation, and reflection within a community context (Bringle and Hatcher 1995). Students provide community service in response to community-identified concerns while they learn about the community context, the connection between their service and their academic coursework, and their roles as citizens. Structured service-learning helps foster civic and social responsibility, is integrated into and enhances the academic curriculum, and includes structured time for students and participants to reflect on the service experience.

Applying service-learning to community GIS as a learning process challenges students to move out of the traditional classroom, beyond their comfort zone, and into the real world among members of a community who may or may not be like themselves. Applying service-learning to community GIS as a teaching method challenges instructors to reenvision the landscape of learning. To frame this new landscape for students means giving them a well-defined set of learning objectives that clearly link the academic context of the course with the community-based experience (Heffernan 2001; Eyler and Giles 1999). These learning objectives should be clear to both students and community partners so that, regardless of the students' learning environment, their learning is focused, contextualized, and assessed. Participants at the Ohio Wesleyan conference specified learning objectives that ranged from "learning about the interaction between geography and diverse disciplines," and "gaining a deeper sense of how history is important in the creation of local social identity," to "developing the ability to design and execute an environmentally and socially based mapping project."[4] Faculty who design courses to include community GIS may consider the following principles of good practice for service-learning pedagogy (Howard 2001):

- Academic credit is for learning, not for service.
- Do not compromise academic rigor.
- Establish learning objectives.
- Establish criteria for the selection of community placements.
- Provide educationally sound learning strategies to harvest the community learning and realize course learning objectives.
- Prepare and train students for learning from the community.
- Minimize the distinction between the student's community learning role and the classroom learning role.
- Rethink the faculty instructional role.
- Be prepared for variation in, and some loss of control with, student learning outcomes.
- Maximize the community responsibility orientation of the course.

Pedagogical challenges are inherent in community-based GIS, and members of several conference teams at Ohio Wesleyan proffered valuable solutions. Strategies included regular meetings to overcome communication gaps, peer mentoring relationships to train students in GIS strategies, and using familiar lab notebooks enhanced with reflective journaling exercises. Students in Macalester College's GIS class, using Ramsey County parcel data to create maps for the planning districts of St. Paul, worked in highly successful learning communities of four to five students and a community partner. Learning groups provided a valuable place to review land-use data, discuss mapping procedures, evaluate the nature of the community, and move closer to articulating district plans.

Students' reflection

Much of what we do with our students in community GIS in a liberal arts setting is experiential. Students move out into the community to prepare maps for nonprofit organizations, historical societies, schools, and government agencies. The students gain new technical skills while they witness GIS in action—GIS helping cities make important decisions about green spaces, plant infestations, health care facilities, recreational trails, and historical sites. However, it is important in these types of service-learning courses to provide opportunities for students to reflect on the service activity to understand and appreciate the intersection between their community-based work, their public contribution, and their academic journey. For one of the project teams, having students develop a deeper understanding of the "productive and humanitarian use of GIS in real-life, real time situations" was an important learning outcome.

To achieve this understanding, reflective exercises for students are critical. Reflection can take many forms, including small learning groups, process meetings in the classroom, journal writing, formal essays, workshops, class discussions, and public presentations. As instructors we need to provide a reflective lens of discovery for our students, helping them to uncover the ethical, social, economic, and political dimensions of the maps they are forging with community partners. National Campus Compact, a national coalition of college and university presidents that promotes service-learning and the civic purposes of higher education, offers guidelines for effective reflection for faculty embedding service experiences into their courses.[5]

The reflective process allows students to engage with the material they are creating for the community in more meaningful ways. For example, one project team noted the illumination that resulted from students' reflection on how to create a map that represented different categories of people as different colors on the map. The process of creating divisions between groups prompted discussion of diversity, provided a profound geographical learning experience, and also made evident "inequities and inequalities in communities" based on the estimated market values of homesteaded parcels.

Community partnerships

Community GIS is dependent on durable, carefully planned, and formally articulated partnerships between the academy and the community. Partnership building is greatly enhanced when both faculty members and community members share community-identified goals and worldviews (Jacoby and Associates 2003; Strand et al. 2003; Gilbert and Sameh 2002). For example, the Swarthmore College team is collaborating with the Chester Consortium for a Creative Community (C4), an organization whose goal to "stimulate problem-solving" with respect to an economically depressed community aligns well with the team's academic goal of applying GIS methodologies to community problems. With the assistance of teams of college students, C4 will realize their vision of creating a history of the city of Chester to demonstrate to younger generations the "precious places" of a once-vibrant city. Students will produce a CD-ROM on the history of Chester for high school students. The CD includes GIS maps created from demographic and geographic data to inform classroom exercises on the structural issues that have "conspired to generate urban poverty."

The sustainability of new partnerships depends on regular communication and time dedicated to relationship building. Participants need to negotiate and to assign tasks, to talk together about parity and inequalities in the relationship, and to build trust. Community partners and university faculty usually come to the relationship accustomed to very different organizational structures, norms, cultures, and professional behaviors and this difference can impact our ability to build alliances. Also, our watches tell very different time. The bells in the ivory tower chime at the beginning and end of a quarter or semester. Community clocks follow grant-funding cycles, nine-to-five work shifts, and seasonal variations. Partnerships need time to grow a culture of mutual respect and communication. When a course includes community GIS, it is essential to welcome community partners into the classroom, to fold students into the partnership, and together formulate questions and process the gathered information. As one project team noted, "face-to-face conversations among students, in a group with the community representative, are highly interactive and valuable. Each side has questions about data, mapping procedures, and the nature of the community."

Successful partnerships are mutually beneficial and reciprocal when the needs of each collaborator are prioritized and resources are shared. CAPHE (The Consortium for the Advancement of Private Higher Education) researchers for the Council of Independent Colleges suggest that community partners may view relationships with colleges and universities from a cost/benefit standpoint (Liederman et al. 2003). Costs of collaboration can include additional work and supervision, use of staff resources, time lost, loss of organizational identity, and a lack of parity. Benefits of partnerships can include the advancement of the group's mission, new perspectives gleaned from students and faculty involvement, access to campus knowledge and resources, grant opportunities, and credibility (Liederman, Furco, Zapf, and Goss 2003). Reviewing the costs and the benefits of the partnership for all stakeholders as the relationship is formed is helpful.

Researchers advise that parity between partners emerges when all stakeholders are focused on a long-term, sustainable relationship that will produce meaningful change for local and global communities (Liederman et al. 2003). Partnerships between the academy and local planning districts, county boards, chambers of commerce, school districts, historical societies, and environmental agencies can benefit by setting up advisory boards and planning teams who share the authority for prioritizing goals and sharing financial resources. Formal partnership agreements that articulate shared goals, strategies, learning outcomes, roles and responsibilities, financial and liability issues, effective communication strategies, and assessment criteria can also help to develop both parity and trust between partners. Some of our project teams have emphatically embraced reciprocity and committed to equal sharing of resources by writing collaborative grants and sharing the costs of GIS equipment and software. For example, the city of Grinnell and two local organizations, Trees Forever (a volunteer-based tree-planting group) and Imagine Grinnell (a grassroots community organization), agreed to contribute funding to purchase new field equipment for the Grinnell College project to define the functions of local city parks, identify areas for bikeway development, and provide tree-mapping inventories. Acquiring a GPS device capable of collecting more precise data will expedite data collection and facilitate the use and maintenance of park maps and inventories. According to the Grinnell College team, "agreeing to share the equipment cost and use has been a big step in solidifying the partnership."

Faculty members developing partnerships and incorporating service-learning pedagogies need not innovate in isolation. Many campuses have centers for community engagement, community service, community-based service-learning, or public service that can (1) assist faculty in brokering new partnerships for GIS collaborations, (2) formulate partnership agreements, (3) provide pedagogical support for courses, and most importantly, (4) ensure that the college or university is ready to "step up" and engage with organizations beyond the context of the GIS partnership.

Sustainable community mapping projects

Campus and community collaborations involving service-learning are a challenge and positive outcomes from such partnerships can be elusive: mapping and GIS add to these challenges.

GIS, GPS, and related technologies are expensive and require specialized knowledge to learn and use. Even if GIS software and hardware are provided by one of the partners (usually a university), it is vital to consider limits on access to the technology (can partners use the technology, when, and where?). Does a software license allow off-campus use or installation of the software on an off-campus computer? Also important are issues of training and use of the software and hardware. What skills do project participants have and are there provisions for training? If skilled GIS users are part of the team, are they responsible for training other participants or actual GIS work or both? What kind of time commitment can they make? If the skilled GIS users are students, will they be available during the summer, on weekends, or evenings (when off-campus partners may be able to meet and work)?

Leveraging future financial support is also critical for the long-term sustainability of participatory community mapping. Financial support may support equipment purchases and fund staff or student help. Many teams at the Ohio Wesleyan conference noted that funding and staffing resources greatly influenced their ability to get projects off the ground. They anticipated the need to secure local and national grants to support their projects. Future collaborations may well depend on the success of pilot projects, such as the ones described in this chapter, to help funders take notice of this important work.

Even with present and future support for GIS technology and skills, it is important not to underestimate the diversity of issues that will affect the positive completion of a collaborative project. GIS data is a good example of an issue that is often poorly understood at the onset of a collaborative project, and this in turn can ultimately undermine the project. Usable data can be expensive and time consuming to acquire. The Ohio Wesleyan Recreational Trails Project owes some of its success to the availability of very detailed and extensive county-level GIS data, maintained and provided by the GIS department in the county. In many mapping and GIS projects, acquiring data comprises the majority of time and expense. To avoid high data acquisition costs, self-collected data is often used. But the costs of technologies used to collect and process such data are not insignificant, nor is the large time commitment. The quality of the data is of fundamental importance: what standards are developed to ensure that data is systematic and appropriately accurate for the given project? The reason that data purchased from private providers is expensive is that good data is expensive to create.

Also necessary to sustain projects is the institutionalization of GIS and related technologies. Many liberal arts schools do not have a geography department. Where, then, will GIS and its associated hardware and software—and the people who maintain and use it—be situated, and who will pay for it? At Otterbein College, GPS technology has found a home in the Education Department and is now an integral part of an educational technology course offering undergraduates the opportunity to teach eighth graders how to apply GIS tools to the preservation of American revolutionary history. Mapping teams of college students and middle school children are using the technology to identify patriot graves in a historical Ohio cemetery as part of a collaboration with the Daughters of the American Revolution and local genealogical societies. This course is part of the Education Department's initiative to become a fully engaged service-learning discipline and will be embedded in the department's curriculum for years to come.

Finally, what are the prospects for institutionalizing GIS off-campus on the sites of our community partners? The recreational trails project at Ohio Wesleyan has benefited greatly in having partners (planners, park managers) who have acquired GIS technology and skills, so they are not wholly dependent on university GIS technology and skills. Universities and colleges need to serve as capacity-builders for these organizations, assisting with training and leveraging future resources to support continued use of these technologies for the common good.

Concluding thoughts

When a campus collaborates with community members, applying GIS to projects of shared interest, we create a complex synthesis of teaching, collaborative work, technology, and real-world engagement. Such projects may seem complex or complicated, but they hold great potential for all participants. Instructors should consider how such projects can reshape their pedagogy: moving beyond closed classrooms, passive lectures, and canned exercises to expose students to collaborative work, reflection, and learning with other students, faculty, and community members. Students learn GIS skills, problem-solving skills, and strategies for effective collaboration. Learning how to collaborate, and learning the benefits of collaboration when solving problems of shared concern, are critical elements of a sound education. Although GIS technology and concepts are often channeled to serve the interests of big business, big government, and big universities, community projects demonstrate how GIS can be used for humane purposes that meet the needs of communities and the mission of liberal arts colleges. The benefits of well-conceived, successful campus and community collaborations using GIS are diverse and important. Take up an attainable challenge and make a difference in the communities surrounding your campus.

Acknowledgments

Thanks to the Midwest Instructional Technology Center (and in particular, Nancy Millichap and Alex Wirth-Cauchon) and Ohio Wesleyan University for support of the conference that inspired this chapter. The Center for Community Engagement at Otterbein College is funded, in part, by the Corporation for National and Community Service, the Learn and Serve Program. The patriot graves project at Otterbein is supported by a grant from Ohio Campus Compact.

Notes

1. See Mapping Campus-Community Collaborations at go.owu.edu/~jbkrygie/comgis/comgis_nitle.html.
2. You can obtain a copy of a blank Action Plan by contacting John Krygier at Ohio Wesleyan University.
3. See teachers.ohiodar.org/cemeteryresearch/otterbein/index.htm.
4. Anonymous quotes from project teams were taken from action plans and progress reports associated with the MITC OWU Conference.
5. See www.compact.org.

References

Aberley, D., ed. 1993. *Boundaries of home: Mapping for local empowerment.* Philadelphia: New Society Publishers.

Bringle, Robert, and Julie Hatcher. 1995. A service-learning curriculum for faculty. *Michigan Journal of Community Service Learning* 2:112–22.

Eyler, Janet, and Dwight Giles. 1999. *Where's the learning in service-learning?* San Francisco: Jossey-Bass.

Gilbert, Melissa, and Catherine Sameh. 2002. Building feminist educational alliances in an urban community. In *Teaching feminist activism: Strategies from the field*, ed. Naples and Bojar, 185–206. New York: Routledge.

Goodchild, Michael. 2002. Forward. In *Community participation and geographic information systems,* ed. W. Craig, T. Harris, and D. Wiener, xxiii. New York: Taylor & Francis.

Heffernan, Kerrissa. 2001. *Fundamentals of service-learning course construction.* Providence, R.I.: Campus Compact.

Howard, Jeffrey. 2001. *Michigan Journal of Community Service-Learning course design workbook.* Ann Arbor, Mich.: University of Michigan Press.

Jacoby, Barbara and Associates. 2003. *Building partnerships for service-learning.* San Francisco: Jossey-Bass.

Kretzmann, J., J. McKnight, and D. Puntenney. 1996. A guide to mapping and mobilizing the economic capacities of local residents. Institute for Policy Research, Northwestern University.

Kwaku Kyem, Peter. 2002. Promoting local community participation in forest management in southern Ghana. In *Community participation and geographic information systems,* ed. Craig, et al., 218–31. London; New York: Taylor & Francis.

Liederman, S., A. Furco, J. Zapf, M. Goss. 2003. *Building partnerships with college campuses: Community perspectives.* CAPHE Publication. The Council of Independent Colleges.

Peluso, N. 1995. Whose woods are these? Counter-mapping forest territories in Kalimantan Indonesia. *Anipode* 27(4): 383–406.

Strand K., S. Marullo, N. Cutforth, R. Stoecker and P. Donohue. 2003. *Community-based research and higher education.* San Francisco: Jossey-Bass.

About the authors

Melissa Kesler Gilbert is the Director of the Center for Community Engagement at Otterbein College where she coordinates community partnerships, service-learning, volunteerism, and community-based action research. She also serves as one of five national Civic Scholars for Campus Compact, appointed in 2003 to examine the intersection of history, civics, and service in higher education.

John Krygier is Associate Professor of Geography at Ohio Wesleyan University, teaching courses in mapping, GIS, and the environment. He recently published *Making Maps: A Visual Guide to Map Design for GIS* (Guilford Press) with Denis Wood, and will be the next editor of the peer-reviewed cartography and GIS journal *Cartographic Perspectives.*

Expanding the social sciences with mapping and GIS

Introduction to chapters 6 – 11

Donald G. Janelle

The social sciences seek to understand human behavior, to predict human actions and their consequences, and to apply such knowledge for resolving human problems. Social scientists have long recognized the significance of place, regional context, and spatial concepts (e.g., distance, location, and proximity) in their theories and models about human interactions. And, although the social sciences have a history of using maps and spatial statistics,[1] the use of spatial reasoning has been constrained by data issues and cumbersome tools. This situation is changing rapidly. Advances in GIS and other spatial technologies have encouraged the use of spatial concepts, enabling empirical analyses of spatial patterns and associations.[2] Now we see a significant interest in formal training in spatial analysis across such disciplines as anthropology, archaeology, economics, history, human geography, political science, and sociology, and in the interdisciplinary areas of criminology, demography, health studies, and urban studies. Increasingly, geocoding (documenting locations) and time-stamping (documenting timing and duration of events) are seen as essential practice in research, enabling scholars to pursue problems in spatio-temporal context to enhance understanding of social processes.

Progressing logically, we extend GIS from research applications into undergraduate education. Technologies have made spatial information readily accessible, via cell phones, GPS units, and digital navigation tools. Software improvements have brought spatial calculations within reach of the undergraduate classroom. Spatial thinking is now being recognized as an essential aspect of contemporary education at all levels. GIS capabilities align with the social scientist's need to explore complex relationships among a large number of interrelated social, economic, and environmental factors. When we introduce students to basic principles and tools of spatial thinking and spatial analysis, we empower them to address issues they may face in meeting career and civic obligations.

Challenges for social science instructors

Traditional social science disciplines and interdisciplinary programs aim to develop a common core of intellectual competencies. As we come to appreciate spatial thinking as part of that core, instructors are finding ways to enhance undergraduate education by applying spatial information technologies to their courses. However, instructors seeking to integrate GIS into courses and curricula face significant challenges. These include:

- Adapting complex software for undergraduate experiences that do not overwhelm the disciplinary content and perspective;
- Applying faculty research to instruction, connecting with students' previous knowledge and curriculum needs;
- Engaging students in proactive experiences of self-learning and group collaboration through research;
- Teaching students to grapple with complexity, to understand the importance of "clean" data, and to account for geographic scale and boundary effects on research outcomes; and
- Teaching students to communicate using maps and other graphic visualizations.

In an era when maps and spatial data are more readily available than ever before, students should learn how to represent information spatially and appropriately. Resolving complex societal issues depends on our ability to decipher intertwined relationships in georeferenced data. For example, crime may be better understood when events are mapped and examined in relation to the neighborhoods where they occur. GIS and the various approaches to spatial analysis also play important roles in attempts to draw associations between human and physical environments. For example, maps of environmental quality and human health can be overlaid to examine correlations.

Meeting the challenges

The chapters in this section illustrate how instructors from several disciplines guide their students in exploratory research and learning, emphasizing principles of scientific problem formulation, data acquisition and measurement, and analysis and interpretation. The authors provide examples of how they challenge students to use GIS to solve problems relevant to the course content. Students also learn how spatial analytic skills can integrate information across knowledge domains, helping to solve problems encountered in political, business, and cultural life. The chapters illustrate solutions to the challenges noted earlier, including such strategies as:

- Live classroom demonstrations of GIS applications;
- Project-based student exercises;
- Group collaboration in problem solving;
- Field research in local areas; and
- Student presentations of research results.

In John Grady's sociology course at Wheaton College, student teams interact as they analyze successive rounds of U.S. census data and explore social diversity in the United States. In the process, they acquire fundamental GIS skills for working with different levels of data aggregation (county, tracts, and blocks), producing maps of social and historical context, and interpreting research results. Linking fieldwork to maps of the study area reinforces the relevance of the spatial data.

At Siena College, James Booker advocates the use of GIS in economics instruction. His students are motivated by an interest in local neighborhoods, and they map interesting data using the Web and a spreadsheet mapping tool. He describes how class discussion is enriched by on-the-spot GIS mapping, addressing students' questions and curiosities about concepts and variables as they interpret economics issues together. Assignments draw on students' familiarity with a region for verification and understanding.

Centre College's A. Endre Nyerges, Daniel Saman, and Laura Whitaker describe how anthropology students are exposed to GIS and mapped data while challenging assumptions about the changing African landscape. Their chapter illustrates how GIS tools can document patterns of change over time to reveal misconceptions about land-use practices in traditional cultures.

Mark Rush and John Blackburn at Washington and Lee University introduce political science students to the role of GIS in redistricting. The potential injustices of gerrymandering motivate students to explore how boundaries affect political power and electoral outcomes. Student teams use GIS to analyze census data and quantitatively assess the fairness of redistricting solutions. They benefit from significant interaction with faculty to resolve technical issues and to gain conceptual understanding.

At Trinity University, Christine Drennon challenged her urban studies students to examine and document the impact of Habitat for Humanity on the San Antonio region. Interdisciplinary student teams used standard GIS processes (variable overlays, address matching, buffering, and neighborhood effects) to measure impacts on the region. They experienced the value of integrating multiple perspectives while solving contemporary social issues.

Pedar Foss and Rebecca Schindler from DePauw University describe how students from four different institutions are working together to develop an Internet-based collaboratory. They are creating a searchable archival database on archaeological surveys in the Mediterranean region. An online seminar on survey archaeology and GIS complements students' experiences as summer interns and work-study students. The seminar features a field-based practicum to design and implement GIS surveys in the students' local areas, engaging them in the full range of issues that archeologists face in landscape investigations.

All of these case studies show how GIS can help prepare the next generation for informed engagement with the complexities of their world.

Notes

1. The Center for Spatially Integrated Social Science at the University of California, Santa Barbara features vignettes on early applications of spatial thinking and spatial analysis by social scientists over the last few centuries in its series of CSISS Classics, at www.csiss.org/classics.
2. A collection of research examples suited for instructors and advanced undergraduate students from across the social sciences is contained in Goodchild and Janelle (2004).

References

Goodchild, M. F. and D. G. Janelle, eds. 2004. *Spatially integrated social science.* New York and London: Oxford University Press.

About the author

Donald G. Janelle is a Research Professor and Program Director for the Center for Spatially Integrated Social Science at the University of California, Santa Barbara, and former Chair of Geography at the University of Western Ontario. His research interests include urban–regional spatial systems development, transportation geography, space–time activity patterns, and the time–geography of cities. Recent coedited books include *Spatially Integrated Social Science* (Oxford University Press, 2004), *WorldMinds: Geographical Perspectives on 100 Problems* (Kluwer Academic, 2004), and *Information, Place and Cyberspace: Issues in Accessibility* (Springer-Verlag 2000).

Professor John Grady is a visual sociologist[1] who was introduced to GIS by a colleague at Wheaton College. Realizing that GIS maps could help students visualize information, he set about mobilizing the Wheaton faculty and administration to embrace the technology as a teaching tool. By January 2003 the college had acquired the facilities and resources for him to offer the half-credit experimental short course described in this chapter. After offering the course three times as a half-credit course, it became a full-credit sociology course in 2007.

To prepare for the course, Grady and Jennifer Lund, an instructional technologist, attended a general four-day GIS training session offered by NITLE during the summer of 2002. Throughout the fall they worked with a librarian to build the basemaps and structure the software skills components of the course. In each course offering, the librarian, T. J. Sondermann, visits the class for an afternoon to teach students about online data resources and how to add that data to their maps. Grady and Lund both provide technology instruction and support to students throughout the course.

Chapter 6

Sociology: Surprise and discovery exploring social diversity

John Grady

Many teachers believe that today's students can be motivated to learn GIS because it is so visual. They hold that lively maps sustain students' enthusiasm as they plow through a software program that has a notoriously long and steep learning curve. I agree that students do find maps interesting and can be readily mobilized to interpret and evaluate them, but they may need more to sustain their interest and engagement. My experience teaching GIS within the context of a sociology course suggests that a deep intellectual encounter may motivate students more powerfully than visual excitement alone. The effectiveness of this approach depends on tight integration throughout the course between the substantive course content and the GIS components, within the framework of a crisply delimited topic and using authentic data from authoritative sources. Students engage with GIS because they seek answers to the compelling questions that arise.

This chapter describes how this strategy is implemented in a two-week, half-credit sociology course, in which students learn to read maps as sociologists, finding correlations and anomalies that inspire hypotheses and further investigation. Here I emphasize activities that relate directly to learning GIS.

In the course, students examine the minority experience in the United States. I begin with demonstrations and exercises that focus on interpreting information, which fosters interest in the subject matter. Rather than treating student curiosity about a map as a potential distraction, which it could easily become, I allot ample time to discuss the maps, examine the distribution of various racial and ethnic minority groups, and share reflections from students' life experiences. My role as instructor is to ground the map in students' own social and historical context and, when possible and appropriate, share and elicit stories relevant to the map. Ideally these stories emerge from a dialogue with students about what they discover upon close reading of the map. Each discussion ends with students listing questions that they wished the map could answer. I point out that they may be able to address these questions as they acquire more GIS skills.

Each map and exercise prepared for this course is intended to show students how and why to use a particular GIS skill. After the introduction of each new skill, students have time to practice that skill while manipulating detailed maps relevant to the substantive focus of the course.

Each map is designed to raise even more questions about the minority experience. Most students quickly develop a deep interest in the topic and its issues, an interest that sustains their motivation to learn GIS skills through their predictable periods of impatience. New skills lead to answers, which lead to new questions, which call for new skills, creating a virtuous circle of surprise and discovery that delights and engrosses our students.

The course structure

The course called Mapping Minorities in the U.S. focuses on investigating patterns of racial, ethnic, and other forms of social diversity. Instruction revolves around a set of GIS maps that depict the United States and Massachusetts. The maps show selected variables from the U.S. Census at the county and the census tract levels. Students also have access to aerial photographs and road maps for several Massachusetts cities. Data for the maps comes from the data library shipped with the GIS software or is downloaded from the American FactFinder Web site maintained by the U.S. Census Bureau (*factfinder.census.gov*).

The course explores the social situation of racial and ethnic minority groups to help students better understand how opportunities to live fuller and richer lives may vary from group to group. This question is important because being a minority, by definition, always entails the possibility of being outnumbered, outvoted, outmaneuvered, shunted aside, or just plain ignored.

Most sociologists believe that racial and ethnic minorities might be better off in contemporary America if they lived in communities where (1) they had access to economic resources, (2) they were socially integrated with the majority population, (3) social relationships within their own groups were stable, and (4) their relationships with other groups were not characterized by marked social distance. Accordingly, the course maps include a number of variables that could be used to measure these factors.

I follow each demonstration of a new GIS skill by an exercise session in which pairs of students practice these skills while researching a question relevant to those four community characteristics. Each team then reports back to the class after a time ranging from thirty minutes to two hours, depending on the complexity of the assignment. The assignments are inspired by a practice used to train medical students in clinical settings and so are termed "GIS rounds." A GIS round includes one short presentation from each team, describing what they found on their maps, what they think it might mean, and enigmas encountered. Teams then field questions from me and their classmates.

Less skilled students use their time mastering the basic GIS skills, while the more adept have time to play with the data and often discover surprising correlations and puzzling anomalies.

Play and exploration is an important pedagogical strategy, enabling students to master skills, generate a need for more skills, and increase their intellectual investment in the series of quests that are the bones of the course. Happily, the discoveries and puzzlements of the more adventurous students both excite and motivate those whose progress is slower, creating valuable teaching moments.

I teach the half-credit course immersion style (from 9 A.M. until 4:30 P.M.) over a two-week period during the January break. During the intervening weekend, students are expected to do fieldwork in a nearby town chosen as a research site. In the first week, students learn GIS and map interpretation skills. In the second week, they apply and hone those skills in an independent research project that culminates in a final report.

Interpreting pattern and variation on a map

On the first day, students learn a set of basic GIS navigation skills. The skills are unified thematically by three issues: discerning pattern and variation in a map, altering scale to change a pattern, and recognizing the importance of place-names on a map.

Pattern and variation

Day 1 begins with a map of the United States that displays concentrations of the four largest racial and ethnic minority groups in the population—Hispanics, African Americans, Asians, and American Indians. The map displays only tracts where the minority group comprises over 30 percent of the total population. Figure 1 shows a varied pattern of racial and ethnic concentration: Hispanics in the Southwest; blacks in the Southeast; American Indians scattered throughout the West, the Great Plains, and the Southwest; and barely visible traces of Asians on the coast of California. The purpose of the map is to introduce the user interface and the basic navigation tools; plus, it simultaneously catapults students into the analysis of racial and ethnic diversity.

Students are fascinated by the map and have many questions about the patterns. Why isn't the Asian presence more visible? Why are the areas of American Indian concentration so large, and why are they often huge rectangular geometries apparently unrelated to each other or to other minority concentrations? Above all, most students are surprised by the apparent geographic separation of concentrated black and Hispanic populations. Our students are mostly white and many are from the northern and eastern parts of the United States. Often students arrive in class assuming that black and Hispanic residents (and, presumably, most other minority groups) live in a sort of jumble in the inner cities. Their curiosity instigates valuable opportunities for discussion, which I exploit by a Socratic line of counter-questioning.

> "Maybe there appears to be more American Indians than Asians because they constitute
> a larger part of the population?"
> "But I thought there were more Asians in the United States?"
> "Actually, you're correct. There are about five times as many Asians as American Indians
> in the U.S.—so it must be something else. What do you think it might be?"

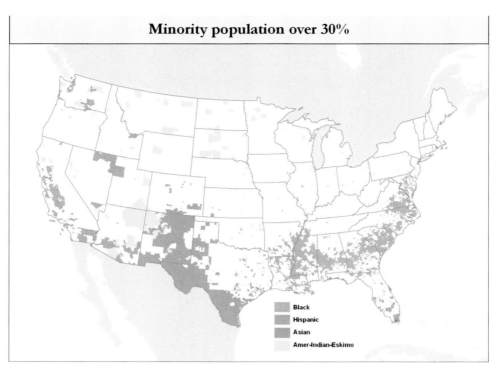

Minority population over 30%

Black
Hispanic
Asian
Amer-Indian-Eskimo

Figure 1. Map of U.S. census tracts in which minority populations exceed 30 percent of total. Colors are 50 percent transparent to reveal overlap.
Data from U.S. Census Bureau and ESRI Data & Maps 2004.

Presented thus as a puzzle, the students become quite animated in their attempts to resolve the question. In only a few minutes I can steer the discussion to a consideration of mapping conventions and the basics of map reading.

"These maps don't tell us how many people live in an area, but only whether a minority group constitutes over 30 percent of the total number of people in the tract."
"Oh, so that means that the Asians might be in small areas with large populations, while American Indians might be in large areas with small populations?"
"Bingo!"

The lessons learned from such a dialogue are made in a context that the students are not likely to forget; they vividly experience the importance of looking at map legends and units of analysis.

This map also provides an opportunity for modeling the use of ethnic history and context. American Indian concentrations are most often found on reservations, an outcome of treaties between the U.S. government and the separate "nations" that previously inhabited the continent and were conquered militarily. Asians tend to be recent immigrants and are often clustered in urban enclaves, especially on the West Coast. Above all, the students are fascinated by the wide geographic expanse, and minimal overlap, of black and Hispanic concentrations in the southern parts of the United States. The same kind of Socratic dialogue leads to a recognition of what was once termed the "black belt" of the slave-owning states in the Confederacy, and currently the area of greatest black concentration. I also lead them to identify an area with high Hispanic concentration that was previously a part of the Mexican Republic, and a Spanish colony before that.

Altering scale

Next the class looks at a very thin ribbon of the Gulf Coast near Biloxi, Mississippi (figure 2). Layers depicting minority concentration reveal virtually no American Indian presence, scattered Hispanic and Asian communities, and a reasonably large representation of blacks. We add a layer of golf courses, recreation areas, hospitals, and the like. Students usually assume that these amenities are indicators of high status—only the prosperous can afford golf courses—and because these resources are reasonably close to where the black population is concentrated, students conclude that blacks in this area must be well off.

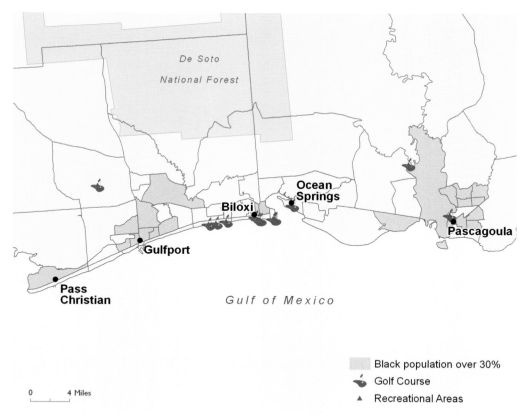

Figure 2. Map of U.S. census tracts along the Mississippi Gulf Coast in which minority populations exceed 30 percent of total.
Data from U.S. Census Bureau and ESRI Data & Maps 2004.

87

We zoom out incrementally until the map includes all of Mississippi and much of Louisiana and Alabama. After each zoom, I ask the students to modify their interpretation of the pattern. By the third or fourth zoom students see New Orleans, and see that the thin ribbon depicting minorities is surrounded by a wide swath with no minority concentrations; other areas of black concentration are about a hundred miles from the coast (figure 3). In contrast to the coastal ribbon with many amenities, these tracts have few if any hospitals or golf courses. Further discussion leads them to propose that, while any of these concentrations might be in areas of minority affluence, it is also possible that the minority groups on the coast might be low-wage service workers in the recreation industry: golf caddies but not golfers; casino workers but not gamblers. They determine they need more information to determine which scenario is more accurate, another important lesson.

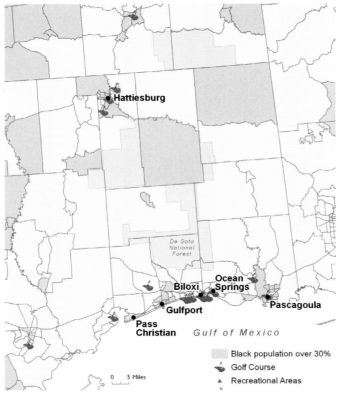

Figure 3. Zoomed-out view of figure 2 map showing census tracts in which minority populations exceed 30 percent of the total.

Data from U.S. Census Bureau and ESRI Data & Maps 2004.

Place and meaning

At the end of the discussion, I reveal that the Mississippi Gulf Coast is well known as a playground for the affluent in the American Southeast. A quick Web search of tourism material shows images of a predominantly white clientele. Students acknowledge that the over-30-percent black tracts are most likely service workers. At this point, we pan over to the island of Martha's Vineyard off the coast of Cape Cod. There is only one census tract on the island with a black concentration over 30 percent.

"Do you think this might be an area of black affluence?"

Having been burned by the Mississippi Gulf Coast experience, they assert that it is probably where the black service workers live.

"Wrong! The census tract is actually in the town of Oak Bluffs and has been a summer vacation area for wealthy blacks for over a century."

The class summarizes these lessons. First, social life is in many ways a narrative of people and events, anchored to place. Knowing the history of a place helps a sociologist understand that narrative and its patterns and devise richer hypotheses to explain them. Second, sound judgments require solid correlations that are always limited by what can be credibly inferred from available data.

Framing the question

On the second day the class works with a map of Massachusetts containing many demographic and socioeconomic variables from the U.S. Census. The Massachusetts map includes ancestry information for close to one hundred different nationalities. In their first assignment of the day, students choose one of three towns near the campus, which will become the research site for their final project.[2] Students use search engines like Google and Yahoo! and databases like the Yellow Pages that might reveal markers of ethnicity in a community. Students look through lists of churches, funeral parlors, restaurants, markets, and other creative locations to find names and addresses that appear to be institutions linked to a particular ethnicity, which might be correlated with areas of minority group concentration. The substantive purpose of this assignment is to demonstrate to students that information for a research project can come from anywhere. A complex society creates rivers of information for a wide variety of purposes that, with some imagination and care, can be trawled for data.

Later in the day, student teams discuss potential research projects with the class as a whole. The other students and I offer comments and coaching to help each team identify interesting and significant questions that are feasible within the available time and that make use of readily available evidence. The discussion accustoms students to thinking and talking about the research process.

The research project

On Day 3 each student team begins building a map of the town that will become its research project. Figures 4a–d illustrate the variation between the towns. Students copy data from the prepared maps to create a general profile of their chosen town, clipping census layers they wish to scrutinize for correlations. In "rounds," students are asked to explain why they have chosen these particular variables as well as the design choices they have made in constructing the layers.

Figure 4a. Location of the three towns in eastern Massachusetts used as research targets. Also, Census tracts in southeastern Massachusetts with percentage of black and Asian populations higher than the state average.

Data from U.S. Census Bureau and ESRI Data & Maps 2004.

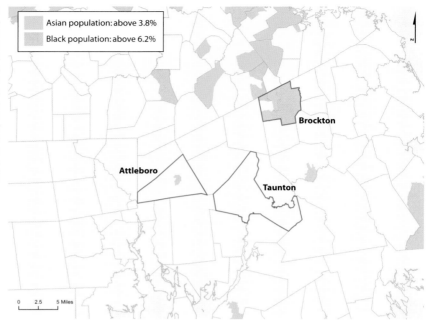

Figure 4b. Census tracts in southeastern Massachusetts where the percentage of people living below the poverty line exceeds the state average, also showing percentage of black and Asian populations higher than the state average.

Data from U.S. Census Bureau and ESRI Data & Maps 2004.

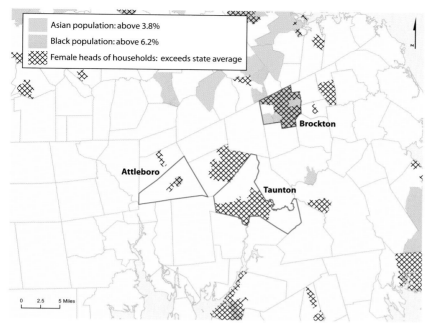

Figure 4c. Census tracts in southeastern Massachusetts with percentage of female heads of household higher than the state average, also showing percentage of black and Asian populations higher than the state average.

Data from U.S. Census Bureau and ESRI Data & Maps 2004.

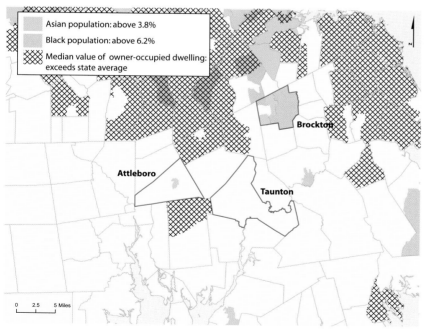

Figure 4d. Census tracts in southeastern Massachusetts with median value of owner-occupied dwellings over the state average. Also, percentage of black and Asian populations higher than the state average.

Data from U.S. Census Bureau and ESRI Data & Maps 2004.

On Day 4 students learn the skills needed to create the layers required by their research hypothesis. Students begin to identify and download variables from the Web and to turn them into layers they can use in their analysis.

Day 5 prepares students to identify, collect, and log relevant data that they will observe when they visit their towns. I deliver a crash course in field work ethics, field note techniques, and use of digital cameras to photograph images for research or display purposes. Students are given a general "shooting script" that acts as a guide to the photographs that best profile a community. In addition, specific instructions are given to each group about what they should look for in pursuit of their particular research question. Students are encouraged to use their imaginations and to photograph and take notes on anything that strikes them as interesting or puzzling. Finally, students are instructed to seek out opportunities to talk with people in the community, explaining their own research interests while soliciting their informants' impressions. In all cases, I instruct students to begin at police headquarters, where they should explain their purpose, deliver the letter of introduction on college letterhead, and ask the police for their observations of and experiences with the community.

For most students this foray into the community is a great adventure and they excitedly relate their encounters the following Monday morning. Typically the interactions seem dramatic only because these students are unaccustomed to sustained conversation with adults from a social context that seems foreign to their own. Students find most people are polite, friendly, and only too willing to talk to them about a topic they feel they know something about. Beyond that, students are excited to see the people and buildings inhabiting places they had previously known only through maps, to find physical artifacts that tell stories about their towns, to discover documentary sources to enrich their town profiles. Students sometimes hear people testify to how much the town has changed since the last census, implying that a literal reading of that data may be misleading; that correlations seen on their census maps may be, if not wrong, at least different than their first impressions. Through this fieldwork students discover the unity of knowledge in a work of their own design, they connect the world of analysis to the world of their perceptions, and they become experts about something in the universe.

Following their field trip, students present a final one-page research proposal, discuss it, and answer questions. Very often they have modified their research hypotheses following their field experience. Most simply sharpen its focus but some alter it more radically. For the rest of the week student teams work independently on their projects. Each day begins with a one-hour session addressing issues of technology, mapmaking, or presentation, contributing to the preparation of their final report. Students put their analysis, maps, and images into a Microsoft PowerPoint presentation. They present their report in an advertised forum on the last day of class, to an audience of students, staff, and a few faculty members in residence during the January break.

Engagement and active learning

One team's research experience provides a clear illustration of the sustaining power of intellectual engagement. Two friends on this team shared an enthusiasm for mapping the sordid underbelly of society. Early in the course they noticed a census tract in nearby Attleboro with a very high concentration of mobile homes. They were delighted to discover what appeared to be a pocket of white poverty in a nearby town. Informed by popular culture and inexperience, they interpreted mobile home ownership to mean low income and perhaps even the "trailer trash" parodied in cult films like Joe Dirt. These two became absorbed with looking for the correlates of "mobile home ownership," eager to expose this apparent squalor that had the dual appeal of being both distinctive from—but also quite close to—their academic ivory tower. They collected their layers and proposed a research hypothesis dealing with poverty and mobile home ownership.

On the aerial photograph of Attleboro, they discovered what appeared to be a large trailer park. It was near the highway, which fit their expectations, and some preliminary map work indicated that this census tract had a high proportion of households on government assistance. With their prurience barely disguised, they set out on their field trip expecting to find residents who were poor, unemployed, and likely to be receiving government assistance.

They returned on Monday excited by their trip, but puzzled and uneasy because they couldn't find the clear evidence of poverty and social disorganization they anticipated. Certainly, it would have been too much to see loud families, trash, and barefoot children—after all it was midwinter—but they were stunned to find that the park was well kept, the mobile homes looked to be in good repair, and, astonishingly, the cars were new, relatively high-priced models in very good condition.

On their return, the students plunged into fevered research, downloading just about every census file they could find about poverty and homeownership. After three days of mapping they realized that what distinguished the mobile home owners in Attleboro from their fellow citizens was not that they were poorer, but that they were older, and significantly so. Fully 80 percent of the mobile home owners were over 55 and more than half were over 75. In addition, most mobile home households had no children and were comprised of either a couple or a single person. By Wednesday, they had their answer. The mobile homes in Attleboro were located in trailer parks and apparently were, for all intents and purposes, retirement communities. The residents could probably afford nice cars because their housing costs were so low. In a perfect example of fruitful research, they tested their hypothesis and found it very wrong, and enthusiastically reported both their discovery and their surprise (figure 5).

93

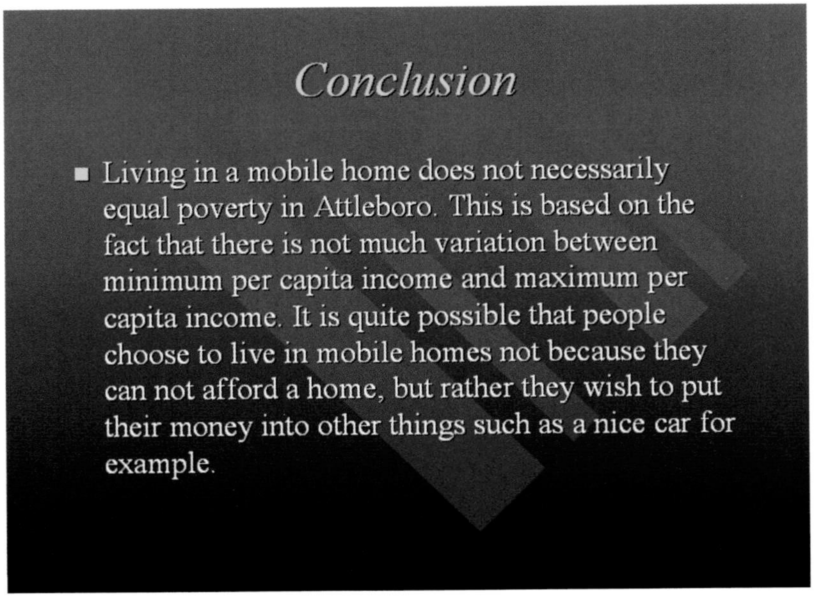

Figure 5. Conclusion of student report on Attleboro mobile home parks (by Kenton Kirby and Michael O'Hara).
Courtesy of Kenton Kirby and Michael O'Hara.

Conclusion

■ Living in a mobile home does not necessarily equal poverty in Attleboro. This is based on the fact that there is not much variation between minimum per capita income and maximum per capita income. It is quite possible that people choose to live in mobile homes not because they can not afford a home, but rather they wish to put their money into other things such as a nice car for example.

The U.S. Census provides valuable information about the circumstances of American lives that can be woven into countless dramas of achievement and loss, integration and separation, focusing on the distinct peculiarities of place. A GIS map populated by census data provides a rich learning environment. Students are enthralled by the questions it raises and the information it provides, as well as engaged by the compelling visual displays. This active learning makes mastery of GIS's forbidding learning curve more palatable, as students absorb the software skills in the pursuit of a genuine curiosity.

Notes

1. The International Visual Sociology Association, www.visualsociology.org.
2. Attleboro, Brockton, and Taunton are all old mill towns with very different histories, economic prospects, and distinct racial and ethnic compositions.

References

Tufte, Edward R. 1997. *Visual and statistical thinking: Displays of evidence for making decisions.* Cheshire, Connecticut: Graphics Press.

About the author

John Grady received a Bachelor of Arts in Asian studies from Boston College, a Master of Arts in Social Anthropology from Yale University, and a PhD in Sociology from Brandeis University. His courses at Wheaton include Society, Technology and the Built Environment; Urban Sociology; Sociology of Work; Sociology of Health and Illness; Global Sociology; Visual Sociology; and Sociological Moviemaking. His research focuses on the influence of the material world of nature and human artifacts on social organization and daily life, the use of visual imagery in social research and analysis, and making documentary films.

To students, economics can seem theoretical and remote, filled with models, equations, and unintelligible graphs. Though our undergraduate students often have part-time jobs and manage some portion of their expenses, few are intimately involved with the broader economy. Within the field of economics, GIS represents a tool with great potential to increase students' understanding and involvement with economics, plus pique their interest in economic theories. When we illustrate the connections between variables, give students the opportunity to link theoretical analysis with projects, and help them visualize the data that they otherwise see only in tabular form, I see them become engaged and eager learners.

In this chapter, I present examples of how we are using GIS across the economics curricula. Maps are applicable to a myriad of topics in introductory economics courses, advanced courses, and student research projects. Given the amount of spatial data with which economists regularly work, it is sometimes surprising to see how little GIS has penetrated into the discipline. I hope my experiences at Siena College and those of my colleagues will provide concrete examples for immediate use and ideas for future projects by both students and faculty.

Chapter 7

Economics: Exploring spatial patterns from the global to the local

James Booker

Economics is a fixture in the undergraduate curriculum. When taught well, students remember examples illustrating economic ideas and are able to visualize the relationships that characterize modern economic life. GIS is one way to bring economic theories to life, and one that can be successfully introduced in lecture, in assignments, and in more involved student projects.

The insights of economics are fundamentally about relationships: between price and quantity, inflation and unemployment, policy and welfare. Ultimately these insights and theory play out and can be explored as numbers and patterns to which GIS is particularly well suited. Seeing the numbers represented through a map may be an important and powerful first step for many. Yet we realize the strength of GIS when we use it as a tool to explore the numbers behind the map. What GIS can best offer economics is access to the data itself, a tool to manipulate data mathematically and iteratively, and a tool to create visual displays all the while. In this chapter I share my own examples and lessons learned, along with stories and examples from colleagues.

Classroom GIS to illustrate economics

Using GIS to display quantitative data on a map gives students a visual perspective that can inspire insights into the economic factors they are studying. Faculty who take the extra steps to visualize data need not have vast expertise with any particular GIS software package, though some do. Many Internet-based mapping applications require no specialized software at all, and more and more federal, state, and regional agencies and offices are sharing their data in this way. Examples include the U.S. Census Bureau online mapping systems,[1] the Demographic Data Viewer,[2] the

U.S. National Atlas mapmaker,[3] and the Bureau of Labor Statistics unemployment mapper.[4] These tools are remarkably accessible to both student and faculty, and the time cost of learning each is measured not in hours and days, but in minutes.

These basic applications hold great promise for improving student learning in economics and in fostering a deeper understanding of the spatial richness of economic systems. They may be particularly useful when the instructor recognizes that visualizing data is a worthwhile activity, but has a limited interest in learning GIS software. Indeed, most economists probably do not even think about spatial visualization as a routinely available option for exploring economic data.

The simplest approach to using GIS in undergraduate economics is to present online mapping applications that demonstrate spatial variability of economic variables. Because introductory economics courses are often large, even at small colleges, opportunities for interactive discussion are limited, and it can be difficult to energize a class about the economic concepts of scarcity, efficiency, and multipliers through lectures alone. The opportunity to visualize "live" data may be particularly important here.

Economist Ken Peterson from South Carolina's Furman University finds that introducing concepts using such "live" GIS presentations sparks student interest. Working directly with a world map and national development indicators from the U.N.'s Human Development Index, he explores basic outcomes such as GDP (gross domestic product) per capita, life expectancy, and population growth (figure 1).[5] He finds the interactive map stimulates students' questions when he can respond immediately to curiosity about, for example, the variation of major macroeconomic variables across nations. Creating maps, during class and in response to student questions, positions him to leverage the "teachable moment," the time when students are most highly motivated to learn.

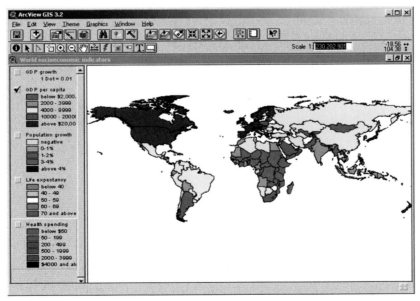

Figure 1. GDP per capita, an example of a GIS view used in lecture during an introductory economics course.
ESRI Data & Maps 2004.

We want our economics students to understand how multiple variables affect an outcome and appreciate the relative importance of each. Glenwood Ross at Georgia's Morehouse College developed a simple GIS project that illustrates the relationship between rail transportation corridors and economic development in Africa. He shows that African transportation networks are dominated by colonial-era projects linking centers of natural resource wealth to port locations (figure 2). Well-developed networks supporting domestic economic activity are limited to South Africa, where many colonizers were also settlers. For Ross, the project "brings life to the words so that my students understand more fully the challenges that inadequate transportation networks pose for economic development."

African Railroads

Figure 2. Rail networks of the African continent.
Map courtesy of Glenwood Ross II, Morehouse College.

Economic geographer Peter Nelson at Middlebury College in Vermont uses GIS so his students can visualize and analyze standard economic concepts such as trade and poverty. He built an example around the siting decisions made by Japanese automakers building new manufacturing facilities in the United States. During a classroom presentation, he points out newly chosen sites and discusses with his students how those decisions were shaped by proximity to transportation corridors, distance from the Midwest hub of the American automobile industry, and the desire to use inexpensive land. Building the map during his lecture, he helps students visualize the multifaceted decision-making process, and he simultaneously introduces basic concepts of spatial analysis. For example, he uses buffers to identify acceptable distances from interstate highways, and queries the data to calculate the number of plants situated within a given distance from the acknowledged center of the automobile industry in Flint, Michigan (figure 3).

Figure 3. Siting Japanese auto manufacturing facilities in the United States illustrates how GIS' analytic capabilities can be used to address economic development.

Map courtesy of Peter Nelson, Middlebury College; additional data courtesy of ESRI Data & Maps 2004.

Class assignments using GIS

In the above examples, faculty themselves direct the visualization experience for students and meet the objective of displaying the data in its various forms. The next step on the continuum is having students themselves manipulate the display of related variables. This may involve a greater commitment, on the part of the faculty and students alike, in terms of learning software. In her environmental economics course at Skidmore College in New York, Melinda Kane asks students to investigate the relationship between population growth and the diversity of federal and state ownership of public land in the western United States. The assignment is based on the "U.S. Public Lands and Population Pressures" data that can be freely downloaded from the ESRI Web site (figure 4).[6] By using an existing, easily available dataset, Kane's students can use the rich visualization capabilities of GIS with minimal startup costs, and Kane's preparation time is also minimized. After the assignment, she presents a lecture on the power of spatial analysis in economics, using examples from her own and others' research.

An alternative approach is to use online mapping systems that minimize startup costs for students and faculty alike. An assignment suitable for students taking a course in economics principles in their first few weeks might compare household income across U.S. states. In this assignment, no in-class instruction is needed. Instead, students are directed to the American FactFinder resource at

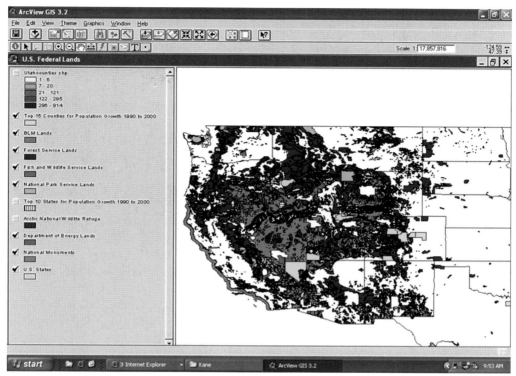

Figure 4. Federal Lands in the western United States is an example of a map prepared in ArcView using free ESRI educational resources available online.
Courtesy of ESRI Data & Maps 2004.

the U.S. Census Bureau Web site, and provided with rudimentary instructions and an appropriate dataset.[7] With this assignment instructors can spend the bulk of their effort and time discussing the reasons underlying the spatial patterns across U.S. states. The ease of implementing this assignment rests not only on the online thematic mapper, but also on the choice of geography: students are very familiar with U.S. states!

In another assignment that minimizes the software-learning thresholds, students used a mapping feature formerly built into Microsoft Office Excel spreadsheet software and still available on many systems.[8] The assignment demonstrated clearly that projects with meaningful spatial content can be undertaken even at a very basic level. I asked students to write short papers in which they would analyze how unemployment rates vary by state. In class I demonstrated the mapping capability in Microsoft Excel, then students calculated the increase in unemployment during the 1990–1991 and 2001 recessions and generated mapped results of their calculations. Although using a very simple tool, with a small set of data, this assignment requires first-semester freshmen to independently pursue the full cycle of inquiry that will be expected of them in higher-level courses, as they find, manipulate, map, and analyze their own data (figure 5).

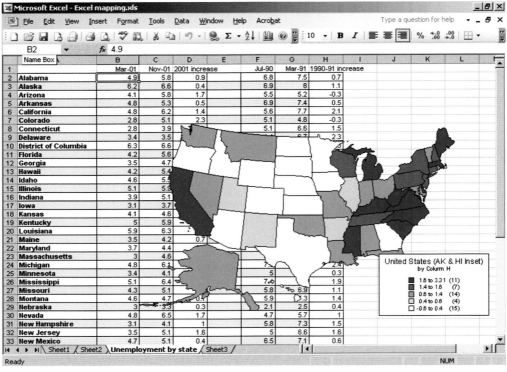

Figure 5. Unemployment by state, created using the Microsoft Excel Mapmaker feature to show variation in unemployment increases during the 2001 recession. This map was created by David Tokarowski, a Siena College freshman in a macroeconomics principles class.

Map courtesy of David Tokarowski.

I have observed another benefit from this assignment. These students are able to write clearly about spatial similarities and differences. In contrast, they are less successful when asked for comparisons across time (e.g., trends in unemployment over the past thirty years), a more traditional perspective for an economist. My experience suggests that teaching students to apply analytic methods in a spatial analysis assignment may be a useful stepping-stone en route to asking them to grapple with time series data. The approach would be consistent with that used in a traditional econometrics course, in which cross-sectional work (using observations across different states, for example) typically precedes time-series analysis.

I have also used an opposite approach with very motivated students in a small economic principles class. In this case my primary aim was to introduce students to the richness and capabilities of professional technologies, and to the availability of specific software tools available on our campus. For this assignment, students compared several measures of income in their home communities with those in New Orleans (figure 6). First, students received a fifty-five-minute introductory lecture and handout on the use of GIS software for the project. They then worked with U.S. Census data[9] to prepare thematic maps at the ZIP Code level showing median income. The assignment was

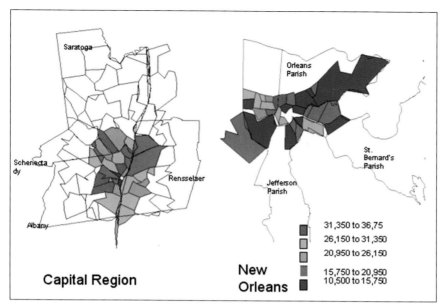

Figure 6. Comparison of household income distributions in a student's home region and the New Orleans area, prior to Hurricane Katrina. This map was created by Siena College sophomore Stephanie VanKempen in a macroeconomics principles class.
Map courtesy of Stephanie VanKempen.

successful in meeting its primary objective, particularly after demonstrating GIS' power in a similar presentation using census block-group-level data. The level of economic analysis in the accompanying student papers was less than in their earlier work; however, it was almost certainly impacted by the additional time needed to learn the basics of the software.

Course-based student projects

We recognize that students benefit from working with and manipulating data directly, and this positive outcome expands when students have first-hand, local connections with the data. In an introductory microeconomics course, Furman University's Ken Peterson is able to link GIS visualization to fieldwork. Using a GIS project prepared by the instructor, students create maps representing poverty and income distribution in the neighborhoods immediately surrounding the university. Students visit the neighborhoods, then use census tract data to show education, race, unemployment, and other variables in preparation for a mock memo to the county council. In this memo, students are asked to identify the extent of poverty in the county, and the characteristics that distinguish rich and poor areas. Peterson found the assignment to be a powerful tool in sensitizing students to differences in economic wealth and income distributions.

Ken Peterson also exploits the motivational power of local regional studies in his urban economics course, where students explore patterns of economic development in neighboring areas. Students use geocoded data from regional planning agencies, including locations and attributes (e.g., sales estimates) of individual businesses. They then test theories such as the economics of agglomeration (which suggests that similar businesses tend to form productive clusters) at an exceptionally detailed level. They learned, for example, that while similar retail businesses often cluster, it is typical to take advantage of traffic traveling in opposite directions. The benefits of actively applying the theories far surpass those of reading about them in a textbook.

At New York's Wells College, students in an environmental and ecological economics course undertake a GIS project focused on the local Cayuga Lake watershed basin. To facilitate the project, economist Kent Klitgaard produced a GIS basemap using the fine-grained long-form census data at the block group scale. In his exercise, students calculate and display income distributions throughout the watershed; later they calculate income distributions for their home counties and then for the nation as a whole. Klitgaard has found that students gain a number of benefits from these exercises: they experience the integration of detailed economic data with spatial analysis, they learn a variety of basic GIS skills, and their interest in the software is heightened by the opportunity to study an area with which they are familiar. When GIS applications cover regions that students can physically visit and explore, the socioeconomic data becomes more "real."

Skidmore College's Robert Jones, also an economist, published a useful description of the wide range of topics appropriate for application of GIS, ranging from international to labor economics, including political economy, public finance, industrial organization, and several others (Jones 2004). As an example, he suggests using tax parcel maps to study public finance. He notes that tax parcel maps are readily available in many areas, as are geocoded county property tax databases. Mapping this publicly accessible data opens up numerous possibilities for straightforward application of hedonic pricing models (figure 7).

But as Jones more fully developed his interests in GIS, he identified many more specific fields within economics in which mapping is useful for students. He suggests:

> At the international level, income levels, growth rates, poverty and inequality, international trade and finance, and the distribution of natural and human resources are prime GIS candidates. At the national level, the location of industries, wage and migration data, regional economic cycles, and studies of the newly defined Economic Areas are topics that lend themselves to GIS analysis. At the local level, urban economic issues such as the dynamics of the relationship between cities and their suburbs, the socioeconomics of neighborhoods and zoning, and tax assessment and environmental planning topics may be addressed by GIS.[10]

Independent student projects and guided research

Undergraduate institutions universally acknowledge the importance of student research opportunities, and economics departments are seeing an increased number of student projects using GIS, ranging from independent study projects and senior theses to research assistantships for faculty or community

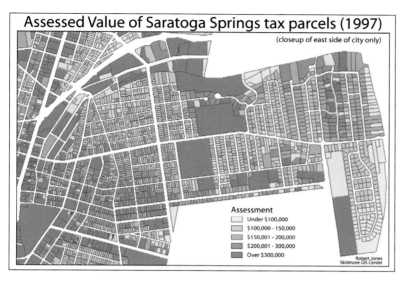

Assessed Value of Saratoga Springs tax parcels (1997)

(closeup of east side of city only)

Assessment
- Under $100,000
- $100,000 - 150,000
- $150,001 - 200,000
- $200,001 - 300,000
- Over $300,000

Robert Jones
Skidmore GIS Center

Figure 7. Tax map used in an introductory GIS class. The wide availability of digital property tax databases allowed visualization of hedonic pricing models in undergraduate economics research.

Courtesy of Saratoga County Real Property Tax Service.

members. Frequently the lines between advanced coursework and research are blurred, and students appreciate the opportunities to work on something "real world" and applied. Russell Patterson, a student at Alfred University in New York, demonstrated the strength of the tool in his independent research on deer hunting patterns in New York State (Patterson, Alsheimer, Booker, and Sinton 1999). Working from county-level data, he constructed trade flows showing the counties where hunters lived and the counties where they hunted. This spatial analysis was a crucial first step in estimating the

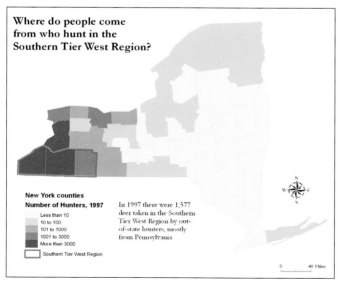

Where do people come from who hunt in the Southern Tier West Region?

New York counties
Number of Hunters, 1997
- Less than 10
- 10 to 100
- 101 to 1000
- 1001 to 3000
- More than 3000
- ☐ Southern Tier West Region

In 1997 there were 1,577 deer taken in the Southern Tier West Region by out-of-state hunters, mostly from Pennsylvania.

0   40 Miles

Figure 8. An example from an undergraduate research project on the economic impact in the Southern Tier West Region from deer hunters originating throughout New York State.

Data courtesy of the New York State Department of Environmental Conservation and ESRI Data & Maps 2004.

regional economic impacts of the fall deer hunting season and demonstrated that counties with the largest deer harvests attract many hunters from local cities, but few from across the state (figure 8).

In addition to the learning opportunities, other important benefits accrue from student projects. Student–faculty collaborations can be rewarding: students benefit both from undertaking challenging applications of technology and from working closely with faculty and participating in the process of economic research. John Buzzard, another student at Alfred University, worked under my supervision to visually define the relationships between economic development, water resource quantity, and water resource quality in the seven Colorado River Basin states (figure 9).[11] This work expanded the understanding of the economic value of water use by illustrating the variation in household incomes and water quantity and quality throughout the watershed.

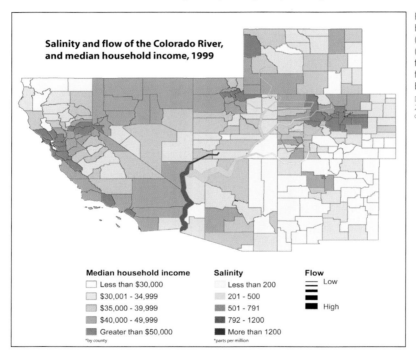

Figure 9. Map relating household income ($/year), salinity (mg/liter), and water flow (shown by width) for the Colorado River Basin states.

Data courtesy ESRI Data & Maps 2004; salinity and flow data courtesy of the author.

Reflections and lessons learned

Despite the promise GIS holds for the undergraduate curriculum, economists have been slow adopters of the technology. Miles Boylan, writing in *Social Sciences Computer Review,* concludes that "economics is clearly laggard when compared to disciplines in the physical sciences, engineering, and mathematics" (Boylan 2004). The relatively slow adoption of GIS as an analytical tool by economists may be related to both the types of questions we typically ask and the types of data we typically use. We have a decided preference for totals: much of the practice of economics is directed toward developing highly aggregated quantitative measures (e.g., GDP), in preference over questions of distributions. GIS is, of course, all about the spatial patterns of distribution. Another inhibiting factor is the discipline's traditional focus on change over time, in preference to change across space. Thus, GIS strongly appeals to scholars asking less traditional questions focused on distribution.

Interest in these less traditional questions is long-standing, although not widely shared. In 1998 a group funded by the National Science Foundation (NSF) studied the potential of enhancing the undergraduate curriculum by incorporating advanced computing technologies and proposed several applications in regional, environmental, and transportation economics and in economics history (Education Center on Computational Science and Engineering 1998). They proposed incorporating discipline-specific software such as econometric tools, GIS development tools such as ArcView, and special-purpose mapping applications such as DDViewer for viewing Census Bureau data. SPACE (Spatial Perspectives for Analysis in Curriculum Enhancement) is a more recent NSF-funded initiative whose scope includes economics. The goal of SPACE is to promote the use of spatial thinking and tools in the undergraduate social sciences curricula, and it provides sample exercises and several syllabi for economics courses.[12] In their workshops, they focus on spatial analysis tools (traditional GIS being only one of many), including GeoDA and GWR, free or low-cost software packages that support spatial statistics and geographically weighted regression.[13] Packages like these address the need by economists and others to statistically explore and validate the patterns that GIS can display so effectively. SPACE itself is a child of CSISS, the Center for Spatially Integrated Social Sciences, which maintains a wealth of resources to support economists interested in spatial analyses.[14] In addition, the GIS initiative of NITLE (National Institute for Technology and Liberal Education)[15] provides a gateway to practices and tools for the liberal arts, and includes a number of resources relevant to the teaching of economics.

Another factor limiting the adoption of GIS may by the discipline's heavy emphasis on theory and generalization, both in research and in the classroom. Yet my experience in the classroom suggests that data and visualization provide a key to involving students with economic theory, particularly in undergraduate teaching. Within the universe of potential GIS technologies, the online mappers are a surprisingly low-cost (in both student and instructor time) approach to introducing data visualization in the teaching of economics.

A tool and data review

The Demographic Data Viewer (DDViewer) is a Java applet (a non-Java version is also available) for basic mapping and statistical calculations of major economic and demographic variables from the 1990 U.S. Census. A major strength of the application is the ease of use and level of spatial detail: data and mapping are available at the state, county, and census-block level. The level of spatial detail makes it possible for students to immediately contrast economic conditions across regions with which they are personally familiar. However, visualization is limited by the tool's inability to perform overlays or utilize more recent data. The Census's own American FactFinder offers similar features in a slightly less convenient interface, but provides a richer and more up-to-date set of U.S. Census datasets. The National Atlas mapmaker uses a convenient interface to show physical and socioeconomic data layers, including average wage per job, median family income, and per capita personal income by year (1979–2003) at the county level. The Bureau of Labor Statistics (BLS) unemployment map utilizes and displays data from the Local Area Unemployment Statistics series. The online tool includes monthly state, county, and Metropolitan Statistical Area (MSA) unemployment data for the years 1978 to the present. The tool is exceptionally easy to use, and maps unemployment levels as well as monthly or annual differences in levels.

Economists are regularly challenged to engage students with their discipline's key concepts. Perhaps GIS, through its ability to humanize topics such as poverty, unemployment, or inflation, can enhance our connections to these issues. We stive to replace sterile charts of guns and butter with more powerful and relevant illustrations of economic principles. GIS can help provide powerful elucidation of change over time and space. An instructor, standing before a class with a detailed interactive map, can encourage questions and respond on the fly, with visual representations of the answers, involving students in the presentation and motivating them to further inquiry.

Current computer technologies make it possible to study the spatial dimension of economics, to expand beyond our traditional analyses of times and totals. Spatial analysis of local markets, trends, and the interplay of local factors offer rich potential for class assignments and student research. GIS tools are well suited to the analytical methods of economics and our students should know these tools for continued study in the field. Regardless of whether they continue in economics, however, they should be exposed to the lessons of regional analysis, learning to think spatially and to consider both the time and the location associated with any information they encounter. As the number of Internet-based mapping applications and readily available spatial data grows, we are slowly removing the barriers created by software, hardware, and lack of expertise that have kept some of us from experimenting with GIS in the past. In the future we will see a range of GIS applications in the economics classroom, far beyond what we see today.

Notes

1. The Bureau of the Census "On-Line Mapping Resources" Web site is at www.census.gov/geo/www/maps/CP_OnLineMapping.htm.

2. You can access the Demographic Data Viewer at plue.sedac.ciesin.org/plue/ddviewer.

3. You can access the U.S. National Atlas mapmaker at nationalatlas.gov/natlas/natlasstart.asp.

4. You can access the Bureau of Labor Statistics unemployment mapper at data.bls.gov/servlet/map.servlet.MapToolServlet?survey=la.

5. This assignment and several others are more fully described in Peterson (2000).

6. See the ESRI GIS in Education site, gis.esri.com/industries/education/arclessons/arclessons.cfm.

7. Map 4 can be produced by selecting "American FactFinder" from the Census Bureau home page, followed by "Maps" and "Thematic Maps." The map is then completed with the choice of variable, such as "median household income."

8. While Microsoft Office 2003 no longer includes the Microsoft Map feature, it remains accessible from the Microsoft Office 2000 or earlier installation CD. Unfortunately, standard Excel installations do not currently support this very simple and useful mapping environment.

9. Summary File 3 (SF3) was used, containing a wealth of household economic data, available down to the subcounty level, and also available by ZIP Code.

10. Robert Jones, personal communication. Jones also maintains a useful Web site at http://www.skidmore .edu/~rjones for his economics and GIS courses, and for GIS resources.

11. Work derived from that shown here was used in Booker (2001).

12. SPACE (Spatial Perspectives for Analysis in Curriculum Enhancement), University of California, Santa Barbara; The Ohio State University; and the University Consortium for Geographic Information Science. csiss.ncgia.ucsb.edu/SPACE.

13. See GeoDa at the Spatial Analysis Laboratory, University of Illinois, Urbana-Champaign (https://www .geoda.uiuc.edu/) and Geographically Weighted Regression (ncg.nuim.ie/ncg/GWR)

14. See the Center for Spatially Integrated Social Sciences at csiss.org.

15. See NITLE resources at gis.nitle.org.

References

Booker, James F. 2001. Increased water demands and limited water resources in the Upper Rio Grande. *The Divining Rod,* New Mexico Water Resources Research Institute. Fall 2001:2–3, 14–15.

Boylan, Miles. 2004. What have we learned from 15 years of supporting the development of innovative teaching technology? *Social Science Computer Review* 22:405–25.

Education Center on Computational Science & Engineering. 1998. Enhancing undergraduate curricula with high performance computing tools and technologies for the California State University System and the National Education Community. Presentation to the Faculty of the Department of Economics, San Diego State University. www.edcenter.sdsu.edu/projects/sdsuecon.html.

Jones, Robert. 2004. Geographic information systems for economists. 57th Annual New York State Economics Association Meeting, Ithaca, N.Y.

Patterson, Russell W., Aaron Alsheimer, James F. Booker, and Diana S. Sinton. 1999. Estimating tourism impacts of deer hunting in New York state: A case study of 1997 deer take data. Proceedings of the 1999 Northeastern Recreation Research Symposium, General Technical Report, Northeastern Forest Experiment Station, U.S. Department of Agriculture.

Peterson, Kenneth, D., Jr. 2000. Using a geographic information system to teach economics. *Journal of Economic Education* 31:169–78.

About the author

James Booker teaches undergraduate economics courses at Siena College in Loudonville, New York. His research interests include water and regional economics and ecological "footprint" analysis. In working with students he strives to illustrate economics in life through the use of real data, believing that the connection between data and theory should occur in many ways, from old-fashioned graph paper, to spreadsheets, to GIS.

GIS and remote sensing can enhance our understanding of the human condition. In particular, these tools help us to examine and compare how people occupy, exploit, and modify the environment across a range of landscapes. During a short, intensive anthropology course at Kentucky's Centre College, students explore the Kilimi area of the Guinea savanna in western Africa and ask questions about spatial patterns. Other anthropologists have previously described the land-use and land-cover patterns in neighboring locations. Armed with the tools of spatial analysis, students study and interpret the Kilimi data to determine whether its human/environment interactions are consistent with these nearby sites. Paper maps and aerial photographs could perhaps be used to achieve this goal, but GIS allows students to simplify the data as a first step in analysis and then ask questions about relationships between data layers. Once they feel confident in their ability to manipulate and understand the data, they move on to more complex, independent analyses. Integrating these techniques doesn't detract from the anthropological content, but rather enhances the ability of students to ask richer questions about the landscape and people's place in it.

Chapter 8

Anthropology: Mapping Guinea savanna ecology in Sierra Leone

A. Endre Nyerges with Daniel M. Saman and Laura B. Whitaker

The notion that Western science misrepresents African reality—whether geographic or otherwise—constitutes a long-standing dogma in the literature. In famous rhymes, for example, the eighteenth-century satirist Jonathan Swift parodied mapmakers who filled unknown spaces with fanciful designs:

> So Geographers in *Afric*-maps,
> With Savage-Pictures fill their Gaps;
> And o'er unhabitable downs
> Place Elephants for want of Towns.
> (Swift 1733, 177–180).

In the 1880s, French explorer Louis-Gustave Binger achieved fame and acclaim as "the new authority" when he demonstrated the nonexistence of the Mountains of Kong that for nearly a century cartographers had mapped running east to west for thousands of miles across interior West Africa (Bassett and Porter 1991 pp. 370, 395–398). A current debate about West African realities, also along the tenth parallel, concerns the effects of human activity on the climatic-vegetation zone known as the Guinea savanna. Alarmed at the loss of wildlife and natural habitat, foresters and conservationists describe a zone of "derived savanna" or grassland created by human-induced deforestation (Nyerges 1987).

By contrast, in their influential *Misreading the African Landscape,* anthropologists James Fairhead and Melissa Leach propose that Western scientists unjustly criticize western Africans by claiming that they are degrading forest to savanna (Fairhead and Leach 1996). Based on studies of unusual vegetation features in the area of Kissidougou, the Republic of Guinea, these authors assert that humans in this zone are in fact creating forest. They describe the Kissidougou landscape as having large (1 to 2 km in diameter), compact, and roughly circular forest islands that surround settlements in an environment of savanna, or wooded grassland. While outside observers have mostly assumed that these patches are the last remnants of a degraded forest, Fairhead and Leach adduce oral and written historical accounts, plus remote-sensing evidence, to contend that Kissidougou is not "half emptied and emptying" of forest, but rather is half full and increasing in forest, under the pressures of human exploitation (Fairhead and Leach 1996, p.2; Overseas Development Administration 1996).

Whether ultimately verified or not, this sophisticated argument forces reexamination of some widely held conceptions of the western African environment. In particular, it raises questions of whether and to what extent the same phenomenon—the creation of forest by rural inhabitants—may be happening in other places; concomitantly, it challenges the general conviction that West African forest is losing ground (Fairhead and Leach 1998). These issues provide the foci for teaching and research questions: How do we appropriately map the human geography of the Guinea savanna? Is the Kissidougou forest islands model applicable to other Guinea savanna areas? If other areas differ, how do they vary, and why? Specifically, how does the location of settlement, identified as the key predictor of forest distribution in Kissidougou, relate to the distribution of forests and other vegetation formations elsewhere in the Guinea savanna? In other locations, do landscape features better predict forest location? Finally, what factors influence the location of settlements, and how do these factors, in turn, relate spatially to forest and other vegetation types?

When addressing issues of the human use and modification of land cover, tools such as GIS and remote sensing effectively complement the core anthropological approaches of ethnography and field ecology. Although few regions of the Guinea savanna are digitally mapped as yet, some suitable material exists to compare Kissidougou to the area of Kilimi, located some 250 kilometers west of Kissidougou in northwestern Sierra Leone (figure 1).

This Kilimi GIS material was developed in the course of interdisciplinary research (Nyerges and Green 2000), in which previous studies of local environmental change dynamics in northwestern Sierra Leone (Nyerges 1988, 1989, and 1992) were integrated with spatial data on the larger area. As an "approach to the analysis of human/environmental interactions," this work proposes "an 'ethnography of landscape', in which the findings of detailed, local ethnographic case studies are combined with remotely sensed information on what the wider region looks like as it changes over time" (Nyerges and Green 2000, p.271).

Out of a commitment to having anthropology students acquire a spatial analytical perspective, we developed a college course on GIS and the Environment that is focused on using GIS as a tool in anthropological research. In this course, students replicate and model the detailed steps of the spatial analysis process that had previously guided research on the Kilimi region, which we have elaborated as a training module for undergraduate students (Nyerges et al. 2005).

This class is taught during Centre's intensive three-week January term, most recently in 2005. During this time, students enroll in one course only. The class meets for four days per week in three-hour sessions devoted to presentations, discussion, and hands-on work, and students are also

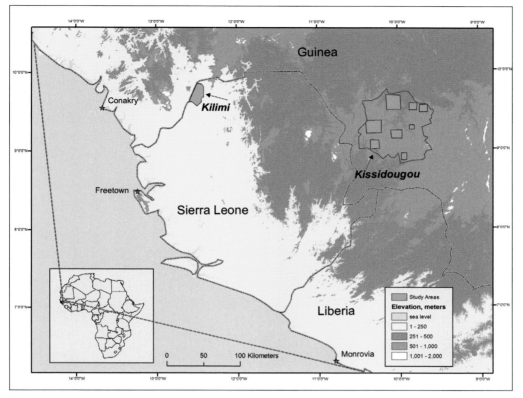

Figure 1. Kilimi and Kissidougou in the western African context. Numbers in the margins denote degrees of latitude and longitude.

Data courtesy of USGS, Shuttle Radar Topography Mission; ESRI Data & Maps 2004.

expected to spend many out-of-class hours on this intensive course. They begin training by taking an online ESRI Virtual Campus course, which guides them in learning the software basics.[1] In our situation, students typically complete the online lessons in the evening and then present and discuss the material the following day. This group learning process greatly enhances the speed with which students are able to grasp the software skills. Thus, every class day students take turns reviewing and demonstrating portions of the course's content that they learned the night before, and their need to articulate and share their understanding of the material with their peers drives them to learn it more thoroughly. The presentations further provide opportunities to correct errors and emphasize key points. Also early in the course, the students consider various examples of GIS use by reading and evaluating chapters from a book on GIS applications in history (Knowles 2002).

Within a few days, the students progress to working with our Kilimi GIS data, and their presentations, demonstrations, and discussions of GIS material are now illustrated by examples relevant to the Kilimi area.[2] For instance, students learn to produce a layout of multiple themes by creating a Kilimi GIS basemap of topographic data.

Examining the streamlines and topographic contours on this map reveals the plain facts of physical geography of the region—that water flows downhill from higher elevations to lower elevations, collecting into fewer, larger watercourses and eventually a river, the Kolenten, at the lowest part of the territory. Further examining the physical geography and settlement locations of Kilimi, as shown on the map, eventually leads students to clarify how people and landscape interrelate in this environment.

Throughout, the class emphasizes the visualization and analysis of spatial relationships, using GIS software to manipulate digital map layers. Students also learn design principles for displaying visual data (Tufte 2001). In another early exercise, for example, students are instructed to generate a layout where all of their Kilimi area data layers, including contour lines, streams, settlements, and vegetation associations, are drawn. The result is a dense and almost unreadable jumble of points, lines, and polygons. They then simplify the image and see how GIS mapping allows them to consider and understand specific spatial relationships. In particular, they learn that by beginning with the topographic basics of settlement, relief, and water flow, they are better prepared to examine, comprehend, and display more complex information about vegetation associations, previously derived from aerial photographs (Nyerges and Green 2000).

By the third and final week of the course, students have acquired sufficient knowledge and expertise to embark on individual projects. Some used GIS to address a research question about the Kilimi landscape: they digitized and analyzed published data on the distribution of nonhuman primates in the Kilimi area (Harding 1984). Others continued to develop materials to teach a particular GIS-based skill using the context of Kilimi data, for example, showing how to "hot-link" digitized slides of the Kilimi environment to appropriate locations on the map. By the conclusion of the intensive course, each student has contributed to an extensive collection of curricular materials—a GIS training module—that can be used to learn about the Kilimi area through a spatial perspective.

This description reflects the version of the GIS and Environment class currently taught at Centre College. By contrast, in an earlier iteration of the course, students spent many hours learning both to produce and manipulate GPS data[3] and to analyze satellite imagery,[4] alongside learning GIS concepts and acquiring ArcView skills. This amount of work, however, resulted in too many late nights in the lab and, worse, ultimately overwhelmed the students with the multiplicity of techniques they needed to master at the expense of the spatial analysis content. We have since modified the course to its current focus on GIS as a tool for anthropological research on the environment.

Spatial analysis of the Kilimi region

Our teaching and research questions focus on the distribution of vegetation formations in Kilimi and include an assessment of whether the patterns are similar to or distinct from those reported for nearby Kissidougou. We concentrate our spatial analyses largely on the interrelated topographic variables that are known to influence vegetation throughout the Guinea savanna: relief, water availability, and the location of settlements. Once students appreciate how these variables are interrelated, illustrated so effectively by GIS, they are then ready to address the interesting ecological and anthropological questions that are the objectives of the course. Consequently, we spend significant

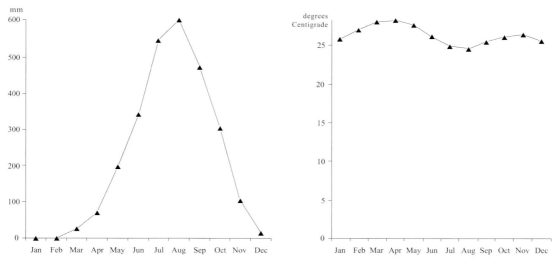

Kilimi Mean Monthly Precipitation

Kilimi Mean Monthly Temperature

Figure 2. Graphs of mean monthly precipitation and temperature in the Kilimi area.
Courtesy of the CLIMATE database version 2.1, W. Cramer, Potsdam, personal communication.

course time learning about Kilimi's climate, physical environment, land-use patterns, and occupants' lives, all the while using GIS-based maps to support our understanding.

Fairhead and Leach open their discussion of the forests of Kissidougou with the short yet memorable phrase: "Kissidougou's landscape is striking" (Fairhead and Leach 1996, p. 1). An appropriate paraphrase for nearby Kilimi might be: "Kilimi's climate is striking." Kilimi's weather is distinguished by a marked alternation between rainy and dry seasons. Although mean monthly temperature remains relatively high and fairly constant throughout the year, rainfall, by contrast, shifts dramatically from near-daily downpours during May through November to the virtual complete absence of rain for much of the rest of the year (figure 2) (Nyerges 1988). Thus, Kilimi vegetation grows abundantly in the rains and then dries and burns ubiquitously in the dry season. Residents also depend on the seasonal rains for farming. Furthermore, surface water in Kilimi fluctuates so that in the rainy season streams are swollen and extensive tracts of land are flooded, while in the dry season even the largest rivers and streambeds are reduced to a trickle of water flow (figure 3).

Kilimi residents modify watercourses only modestly, if at all, and they depend on existing streams for all their water needs including drinking, cooking, washing, bathing, and building. The task of collecting water and carrying the heavy, open basins home over uneven terrain is common daily work, especially for women and many children. Because they require accessible water, even during the dry season, villagers prefer to situate settlements near permanent streams (figure 3), and, as GIS analysis readily shows, virtually all Kilimi villages are located within 600 meters of mapped streams. As students consider these distinct map layers, in which villages are encoded as points and streams and elevation contours as lines, the critical and enduring relationships between settlement location,

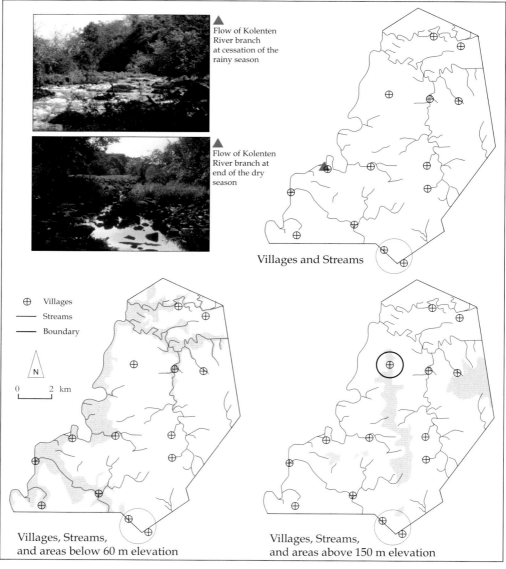

Figure 3. Seasonal flow of the Kolenten River and simplified maps of Kilimi basemap features. (1) Villages and Streams—Kilimi settlements are tied to streams where water flows throughout the year. (Note that the two southernmost villages, circled in gray, are located on streams off the mapped area.) (2) Villages, Streams, and areas below 60 m elevation—Most Kilimi settlements are found at or below this elevation, where groundwater discharge from higher zones helps maintain water flow through the dry season. (3) Villages, Streams, and areas above 150 m elevation—Only one Kilimi village, circled in black, is more strongly associated with higher elevations than with mapped streams. The village is likely the oldest Kilimi settlement, and its location was perhaps determined by conditions general to Sierra Leone during the nineteenth century. At that time, almost constant warfare associated with slave raiding led to a system of wartowns that were situated on defensible high ground and surrounded by a living wall of trees (Atherton 1979, pp. 36–70).

Photos by A. Endre Nyerges; topographic data available from USGS.

water flow, and elevation in this environment become readily apparent to them. In brief, the images they study illustrate a major consequence of hydrological conditions for Kilimi residents—that permanent settlements are "tethered" to the locations of year-around water flow.[5]

Once students have a firm grasp of the interrelationships among these topographic variables, we turn our attention to vegetation patterns. The previous research on which the course is based has shown that the vegetation formations of the Kilimi area include forest, defined as closed-canopy tree cover with no grass in the understory; hardpan, a treeless savanna of short grass, 1 meter tall, on lateritic rock; boli, inundated grassland on low-lying areas of poor drainage; and savanna wood-lands, a "residual" category of tall grasses, 2 to 4 meters in height, interspersed with open-canopy fire-resistant trees (Nyerges and Green 2000; Tegler 1983).

Students can readily employ GIS to analyze the data layers depicting the Kilimi area vegetation, showing that these formations are composed of 8 percent forest, 14 percent hardpan, 4 percent boli, and 74 percent savanna woodlands. Yet once they overlay the vegetation data layers with the others in the Kilimi basemap, they find that the actual distribution of land-cover types in relation to topography is complex and decidedly difficult to interpret. Distinctively nonrandom patterns begin to emerge from the data only when students isolate specific features, such as forests and streams, and display them together in ArcView (figure 4).

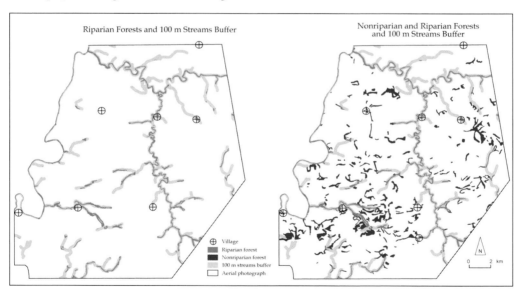

Figure 4. Forest patches are widespread across the Kilimi landscape, but GIS analysis clarifies some of the nonrandom features of this distribution. (1) Riparian forests and 100 m Streams Buffer—Forest covers much of the area adjacent to mapped streamlines; indeed, while the 100-m streams buffer covers 12.4 percent of the Kilimi aerial photo, this area includes a disproportionate 34.2 percent of the Kilimi total forest. (2) Nonriparian and Riparian Forests and 100-m Streams Buffer—The remaining 65.8 percent of the Kilimi forest is scattered over the rest (87.6 percent) of the landscape outside the 100-m streams buffer.

Topographic data available from USGS.

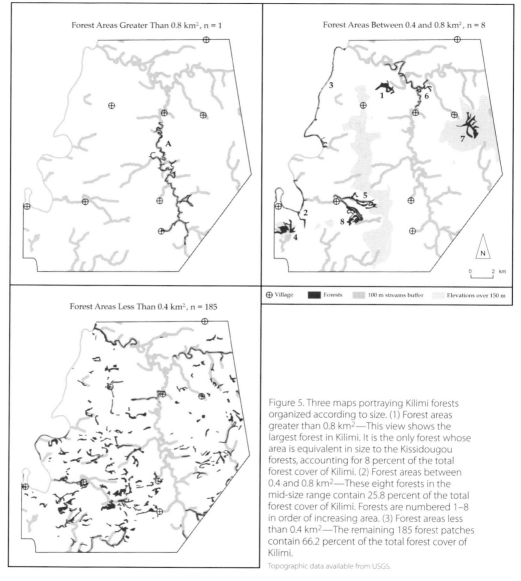

Figure 5. Three maps portraying Kilimi forests organized according to size. (1) Forest areas greater than 0.8 km² —This view shows the largest forest in Kilimi. It is the only forest whose area is equivalent in size to the Kissidougou forests, accounting for 8 percent of the total forest cover of Kilimi. (2) Forest areas between 0.4 and 0.8 km² —These eight forests in the mid-size range contain 25.8 percent of the total forest cover of Kilimi. Forests are numbered 1–8 in order of increasing area. (3) Forest areas less than 0.4 km² —The remaining 185 forest patches contain 66.2 percent of the total forest cover of Kilimi.

Topographic data available from USGS.

Employing the geoprocessing extension of this program enables the delineation of a 100-meter-wide buffer zone around mapped streams, thereby defining a category of riparian forests. This simplified and focused mapping reveals clearly that, unlike the peri-village[6] clustering of trees characteristic of Kissidougou, many Kilimi forests are distributed alongside streams. As with settlements, the single most important factor determining the location of forest area in Kilimi is the presence of flowing water.

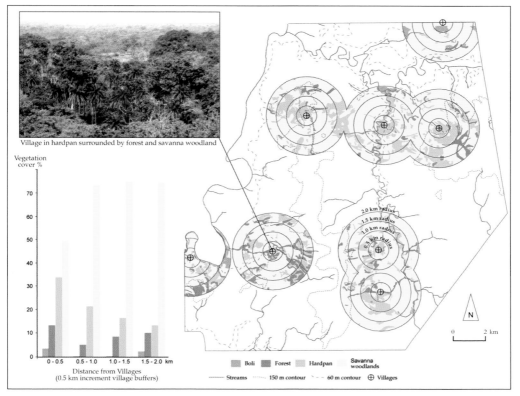

Figure 6. Photograph, map, and graph illustrating the distribution and variation of vegetation formations in proximity to eight Kilimi villages. The photograph shows a distant view of a Kilimi village and its typical surrounding vegetation. On the map, 0.5-km incremental buffer rings encircle each village, and the buffer rings portray vegetation distributions up to a 2-km radius distance from village sites. The accompanying graph quantifies the percentage of boli, forest, hardpan, and savanna woodlands cover in the buffer rings.

Photo by A. Endre Nyerges; topographic data available from USGS.

The class then takes its analysis of Kilimi forests one step further by separating the forest patches into three size classes (figure 5). For nearby Kissidougou, Fairhead and Leach describe the forest islands as 1 to 2 kilometers in diameter, compact, and roughly circular in shape, suggesting forests of 0.79 to 3.14 kilometers^2 in size. In contrast, students find that only one of the 194 forest patches in Kilimi (figure 5) falls into this Kissidougou size range. Unlike the compact and roughly circular peri-village forests of Kissidougou, this forest patch is long and sinuous, and it obviously follows a river. Other, somewhat smaller, forest patches in Kilimi are also clearly riparian, or, in some cases, they are associated with elevations of 150 meters and higher. In each instance the location or shape of these forests is better explained by the presence of water or a location at higher elevation, rather than by proximity to a settlement.

Finally, many of the 185 patches in the smallest forest size category are fragmentary and also adjacent to mapped streams. Alternatively, their characteristically thin and curvilinear shapes suggest the presence of unmapped, seasonal streams where groundwater may be close enough to the surface to be tapped by tree roots in the dry season. Most importantly, this predominance of small forest fragments in Kilimi shows that the Kissidougou model of large forest islands is not necessarily applicable throughout the western African Guinea savanna. At this point, students can use their newly expanded understanding to argue that the location of water, rather than settlement, is the critical determining factor in the occurrence and character of forests in the Kilimi landscape.

Using GIS, students demonstrate that the Kissidougou forest islands model per se does not apply in Kilimi, and they soon find that the larger vegetation patterns surrounding villages are also distinctive. By buffering the areas around each village, students can isolate and quantify the land uses and vegetation associations in near-village zones, and they realize that people in the Kilimi environment settle in the most diverse locations, presumably to access their varied resources. For example, in Kissidougou the area up to 0.5–1.0 kilometer radius around settlements is entirely forest-filled. By contrast, the equivalent area around Kilimi villages consists of a mixture of hardpan, forest, and savanna woodlands, and forest cover does not dominate (figure 6). Indeed, land nearest a Kilimi village may even include boli if the settlement in question is specifically exploiting this resource. In general, the observed mix of vegetation types in near-village lands clearly confirms the argument that Kilimi residents deliberately choose village sites that allow them the best access to a variety of resources.

Fairhead and Leach attribute the presence of forest islands around Kissidougou villages to residents' activities that modify and improve conditions to stimulate the growth and reproduction of trees (Fairhead and Leach 1996). Likewise in Kilimi, villagers' activities may stimulate forest growth. For example, residents may plant economically valuable trees, such as banana, mango, lemon, orange, papaya, and kola nut, and they may deliberately encourage oil palms and cotton trees. The tending of kitchen gardens may increase the quality and fertility of soil in the village area as soil is loosened and heaped for cassava mounds, and kitchen ash and waste are discarded nearby.

Additionally, the life cycle of housing might contribute to longer-term patterns of forestation. The structures, built with thatch or grass for roofs, wattle or poles tied together with grass or vine for frames, and daub or mud for walls, are relatively easy and inexpensive to make, but are not likely to last beyond five to ten years. When it is termite-infested and no longer keeps out the rain, the house is abandoned to ruin and is replaced with a similar structure. Ultimately, as happens all too often in this environment, dry season bush fires may ignite the thatch roofs of houses and whole villages may burn (figure 7). If a village burns late in the dry season, it may no longer be possible to gather any remaining grass to serve as roofing material. A village that has suffered such a fate may be abandoned for years while its residents dwell elsewhere, pending a future return. During this time, forest may grow on the site that is now well fertilized and improved.

Nevertheless, the distribution of Kilimi forests is better explained by stream locations and water availability than by village locations or human actions alone. To confirm this association, students use GIS to compare the proportional amount of forest cover in Kilimi areas both within village buffers and in proximity to streams (table 1).

The wattle-and-daub of construction of houses.

Surveying the smoldering village just minutes after a fire consumed half the residences.

Figure 7. Kilimi villages are small concentrations of biomass and fertilizing activity on the landscape, as these pictures portray.
Photos by A. Endre Nyerges.

Thatch-roofed structures of a Kilimi village.

Table 1. Kilimi forest areas in proximity to villages and streams.

Kilimi village and stream buffer areas	Percentage of Kilimi forest area within buffers
0.5 km village buffer—2.7% of Kilimi land area	4.5%
20 m stream buffer—2.6% of Kilimi land area	14.5%
1 km village buffer—10.6% of Kilimi land area	9.3%
80 m stream buffer—10.0% of Kilimi land area	32.4%

As the results show, the Kilimi areas defined by vicinity to streams are always more forest-filled than areas of equivalent extent defined by proximity to villages, by a factor of 3.2 to 3.5. Although both villages and watercourses may predict the occurrence of forests in this Guinea savanna environment, water location is by far the better predictor. This finding contrasts with the Kissidougou case where settlement and human action seemingly determine forest location.

Reflections and lessons learned

By applying GIS tools and techniques in this short, intensive anthropology course, students learn to visualize the Kilimi landscape and to ask and answer questions about spatial patterns embedded in the geographic data. More traditional representations of spatial data, such as paper maps and aerial photographs, could perhaps be used to achieve this goal, but GIS allows students first to simplify and study the data layers individually, and then to group them in combinations that address questions about human/environmental relationships across data layers. Acquiring expertise with GIS can be time-consuming, but we design that process to support the course's analytical goals by having students learn the GIS they need to know with the Kilimi data. Once students acquire expertise in manipulating spatial data and better understand the landscape they are studying, they have confidence to generate new data and analyses through their independent projects. Students find the work both challenging and worthwhile, and they profit by gaining an appreciation for how spatial analysis can be applied to other topics. For example, Daniel Saman is eager to study the patterns of cancer as they may relate to environmental variables.

How do we appropriately map the human geography of the Guinea savanna? On August 23, 1796, in the course of explorations on the flow of the Niger River, Scottish explorer Mungo Park climbed a hill in West Africa and observed in the distance some mountains, which he was told were situated in the Kingdom of Kong (Bassett and Porter 1991). There was no reason to doubt Park's observations at the time, and it took almost a century for another visitor, this time armed with a sextant, to erase the error of extrapolating a few small promontories into a vast mountain barrier extending across half the continent. Although the peri-village forests of Kissidougou are hardly in the category of the Mountains of Kong, the research and class work described here nevertheless illustrate an important value of GIS in anthropological research: to test concepts of human–environment interactions generated for one area against the differing, but equally valid, realities of other areas.

Acknowledgments

Writing this paper was supported by a summer research grant from the Centre College Faculty Development Committee (FDC). Our work on the project was greatly enhanced by the facilities of the newly housed and refurnished Centre Parents Association Social Sciences Laboratory and the Arthur Vining Davis Foundations High Performance Computing Center (HPCC). ESRI, through its University Relations program, donated the ArcView licenses for our lab, while Centre's Information Technology Service (ITS), and Keeta Martin in particular, gave timely technical assistance. Lee Blonder, Diana Sinton, Jenni Lund, and an anonymous reviewer provided valuable editorial comments, and Robert S. O. Harding supplied archives. Diana Sinton drew the map in figure 1, and Michael Law drew the maps in figures 3, 4, 5, and 6. Endre Nyerges's original training in spatial analysis was funded in part by an NSF Scholars Award in Methodological Training for Cultural Anthropologists (NSF SBR 9615826).

Notes

1. We use ESRI ArcView 3.2 for this course, and students complete the Virtual Campus Introduction to ArcView 3.x online lessons. Students can log on in the lab or download an ArcView training version that comes with the course registration and work on their own machines. For several years, the ESRI higher education team has donated Virtual Campus registrations to the students.
2. Their reading at this time covers selected aspects of the controversy—over the processes and patterns of forest ecological change in western Africa—that generated the original research.
3. Using a Trimble GX Pro XRS Receiver with TSC1 Asset Surveyer.
4. Using Landsat satellite images of western Africa and the MultiSpec software.
5. According to the concept of resource "tethers," if a resource is limited in distribution then the location of settlements that depend on the resource may be concomitantly restricted, to the extent that available technology does not overcome the limitations (Sutton and Anderson 2004, pp. 43–44).
6. Encircling the perimeter of a village.

References

Atherton, John H. 1979. Early economies of Sierra Leone and Liberia: Archaeological and historical reflections. In *Essays on the economic anthropology of Liberia and Sierra Leone.* Liberian studies monograph series, number 6, eds. Vernon R. Dorjahn and Barry L. Isaac, 27–43. Philadelphia: Institute for Liberian Studies.

Bassett, Thomas J., and Philip Porter. 1991. "From the best authorities": The Mountains of Kong in the cartography of west Africa. *Journal of African History* 32:367–413.

Fairhead, James, and Melissa Leach. 1996. *Misreading the African Landscape—Society and ecology in a forest-savanna mosaic.* African Studies Series, 90. Cambridge: Cambridge University Press.

———. 1998. *Reframing deforestation—Global analysis and local realities: Studies in west Africa.* London: Routledge.

Harding, Robert S. O. 1984. Primates of the Kilimi area, Northwestern Sierra Leone. *Folia Primatologica* 42:96–114.

Knowles, Anne Kelly. 2002. *Past time, past place: GIS for history.* Redlands, Calif.: ESRI Press.

Nyerges, A. Endre. 1987. The development potential of the Guinea Savanna: Social and ecological constraints in the West African "Middle Belt." In *Lands at risk in the third world—Local-level perspectives,* ed. Peter D. Little and Michael M Horowitz, with A. Endre Nyerges, 316–36. Institute for Development Anthropology, Monographs in Development Anthropology. Boulder, Colo.: Westview Press.

———. 1988. Seasonal constraints in the Guinea savanna: Susu ecology in Sierra Leone. In *Coping with seasonal constraints,* ed. Rebecca Huss-Ashmore, with John J. Curry and Robert K. Hitchcock. Special issue. *MASCA Research Papers in Science and Archaeology* 5:86–95.

———. 1989. Coppice swidden fallows in tropical deciduous forest: Biological, technological, and sociocultural determinants of secondary forest succession. *Human Ecology* 17(4): 379–400.

———. 1992. The ecology of wealth-in-people: Agriculture, settlement, and society on the perpetual frontier. *American Anthropologist* 94(4): 860–81.

Nyerges, A. Endre, and Glen Martin Green. 2000. The ethnography of landscape: GIS and remote sensing in the study of forest change in west African Guinea savanna. *American Anthropologist* 102:271–89.

Nyerges, A. Endre, Ryan E. Bowe, G. Kelly Kron, Daniel M. Saman, and Laura B. Whitaker. 2005. Spatial analysis of Kilimi, Sierra Leone: A GIS module for training and research. Centre College, Danville, Kentucky.

Overseas Development Administration. 1996. *Second nature: Building forests in west Africa's savannas.* Video-cassette. Hayward's Heath, U.K.: Cyrus Productions, Ltd.

Sutton, Mark Q., and E. N. Anderson. 2004. *Introduction to cultural ecology.* Walnut Creek: Altimira.

Swift, Jonathan. 1733. *On poetry: A rhapsody.*

Tegler, Brent. 1983. Environment and plant associations in Kilimi, Tambakha Chiefdom, Sierra Leone. Report submitted to the Park Director, Outamba-Kilimi National Park, Ministry of Agriculture and Forestry, Sierra Leone, World Wildlife Fund–U.S.

Tufte, Edward R. 2001. *The visual display of quantitative information.* 2nd ed. Cheshire, Conn.: Graphics Press.

About the authors

A. Endre Nyerges is an associate professor of anthropology at Centre College in Danville, Kentucky, where he teaches courses in physical anthropology and archaeology, human evolution, GIS, ecological anthropology, environmental studies, peoples of sub-Saharan Africa, and classics of ethnography. He has applied spatial analysis tools such as GIS and remote sensing to his ethnological and ecological research in western Africa for many years.

Daniel M. Saman (class of 2006) and Laura B. Whitaker (class of 2007) are both undergraduate students at Centre, majoring in anthropology. They were members of the January 2005 session of Nyerges's GIS and the Environment course and subsequently worked with him to further develop curricular materials during the summer of 2005.

Politics 295: Seminar on Technology and Politics: Redistricting and Beyond is an experimental course at Virginia's Washington and Lee University in which we apply GIS to an important political phenomenon: the redistricting process. Our goal is to teach students to use GIS technology in a politics laboratory. Through maps, students work with large datasets that represent voting patterns in legislative districts and propose changes to these districts' boundaries that reflect statewide demographic trends. Throughout the iterative process of generating new voting districts, they justify their reasoning and conclusions visually with new maps.

Merging mapping technology into the course has been a worthwhile but complex process. Political scientist Mark Rush worked closely with John Blackburn, Washington and Lee's director of instructional technology, to plan and execute the various steps. Neither of them had worked extensively with GIS software before this course, and they found that integrating these technologies required time, patience, time, forethought, luck, and then some more time. But they have learned a lot and will do it again soon.

Chapter 9

Political science: Redistricting for justice and power

Mark Rush and John Blackburn

The political science classroom is an ideal venue for introducing quantitative analytical methods. Grounded in the vast, quantitative, numerical datasets represented by the decennial U.S. Census, studies of reapportionment and redistricting merge the field of statistics into political analysis. In turn, political analysis has an intrinsic appeal because it engages students in quantitative topics through their natural interests in national and international current affairs. Fueled by interest in political power and control, students enthusiastically acquire, manipulate, and analyze data to test hypotheses and draw conclusions. As with other methods courses, when we demonstrate the use and relevance of the analytical tools and procedures to topics that appeal to the students, their interest in and deep understanding of the material is that much greater.

Our course focuses on the decennial redistricting process, where voting patterns, electoral law, voting rights law, and Supreme Court jurisprudence intersect. With a learned appreciation for the vast literature on the subject, students use GIS technology to redraw our state's legislative or congressional district boundaries, under the same legal constraints imposed on government officials who engage in a parallel process.

In this chapter, we briefly discuss the legal and academic background of the redistricting process. We then turn to the teaching issues and logistical administration of a course that integrates GIS instruction with lectures on the law and politics of redistricting.

Census and decennial reapportionment

From the beginning, the United States wanted to ensure that all its citizens have representation in the federal government, and thus created the 435-member House of Representatives. Article 1, Section 2, of the United States Constitution states:

> Representatives and direct Taxes shall be apportioned among the several States which may be included within this Union, according to their respective Numbers . . . . The actual Enumeration shall be made within three Years after the first Meeting of the Congress of the United States, and within every subsequent Term of ten Years, in such Manner as they shall by Law direct.

That is, since our population is in a constant state of flux (through immigration, emigration, and moves by people within the country), we need to count our population every ten years, to know both how many people are here and where they live: thus we hold a nationwide census once each decade. When that count is complete, the House of Representatives undergoes reapportionment, the process by which the 435 total seats are reallocated among the states to account for the population shifts of the preceding decade. Since all states are entitled to at least one seat in the House, the allocation of seats is not exactly proportional because some states, such as Wyoming and the Dakotas, have very small populations. But, once congressional seats are apportioned among the states,[1] the state legislatures must then arrange to redraw congressional and legislative district boundaries to balance their populations.

Redistricting is the process by which legislative district lines are actually redrawn—by partisan actors such as state legislators, by nonpartisan or bipartisan actors such as redistricting commissions, or possibly by judges. Gerrymandering, an informal term that refers to the use of the redistricting process to achieve (or enhance the likelihood of achieving) a particular electoral result, is done by drawing districts in a manner that either enhances or dilutes the voting strength of particular blocks of voters.

Gerrymandering typically refers to bizarrely shaped legislative districts that are drawn in a manner that snakes around, picking up voters here and avoiding voters there. The term was

Figure 1. The Massachusetts gerrymander of 1812. While the district was drawn by Jeffersonian Republicans who hoped to dilute Federalist voting strength, it was branded a gerrymander, both because it resembled a mythical salamander and because the districting plan had been approved by then Massachusetts Governor Elbridge Gerry.
Elkanah Tisdale, "The Gerrymander," Boston Gazette, March 26, 1812.

coined in 1812 by a political cartoonist in Massachusetts in response to the state's newly drawn legislative districts (figure 1). In the previous election, the Jeffersonian Republicans had gained control of the legislature and used their majority power to redraw the state legislative districts to enhance their electoral prospects. One such district, on the north shore of Boston, was so contorted that a cartoonist said that if it had wings and claws, it would look like the mythic salamander. Since Massachusetts Governor Elbridge Gerry signed the districting plan into law, the plan was deemed a "gerrymander" even though he was not its perpetrator and the form was designed, in fact, to do harm to the electoral prospects of his own Federalist party.[2]

The current context: legal and constitutional constraints

Myriad legal and constitutional provisions at both the state and federal level constrain how we can draw legislative and congressional districts. The history of voting rights law has many odd twists and turns, many of which are beyond the scope of this chapter. Currently, however, two particular constraints frequently impact the drawing of district lines: equal population and minority voting rights.

Equal population

For all intents and purposes, legislative and congressional districts must be comprised of equal numbers of people. The so-called "one-person, one-vote" rule was first set forth by the Supreme Court in Baker v. Carr (1962) and Reynolds v. Sims (1964) where the court ruled that state legislative districts had to have equal populations.[3] Court cases have brought the amount of variation of these "equal" numbers of people to less than 1 percent in congressional districts.[4] Over time, however, the court has loosened the rule for state legislative districts to facilitate the preservation of existing municipal boundaries, communities of interest, and so forth, and the number of people in each district varies more, by up to 10 percent in some situations.[5]

Minority voting rights

Following the 1965 Voting Rights Act and its subsequent amendments, states are required to take minority representation into account when drawing district lines. This act and its amendments led to extensive litigation in the 1990s that focused on balancing the competing claims of the Voting Rights Act (which instructed states to take group voting rights into account when drawing district lines) and the Equal Protection Clause of the 14th Amendment (which a majority of the Supreme Court read as forbidding the use of race as a criterion for public policy). Recently, in Easley v. Cromartie,[6] the Supreme Court finally set forth a rule that allowed states to foster minority representation rights while not running afoul of the 14th Amendment. Despite this ruling, expensive redistricting litigation continues to flourish because so many different political, advocacy, and special interest groups seek to ensure that redrawn district lines do not threaten their political fortunes or agendas.

Many people associate gerrymandering with bizarrely shaped districts, but a district's shape is not a particularly important factor in redistricting analysis. In fact, GIS technology enables those with a particular agenda to create districts that dilute or enhance the political power without necessarily having to draw a bizarre district, though sometimes they do emerge. Oddly-shaped districts often arise as politicians struggle to meet the legal and constitutional constraints described earlier. The Virginia state legislative plan was challenged in court because some of its districts, drawn to protect minority voting interests, are indeed bizarre. Virginia

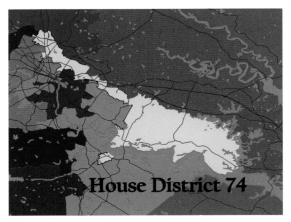

Figure 2. Virginia House District 74, known as the "hatchet" district for its distinctive shape.

District boundary data from U.S. Census Bureau. Road and stream data from ESRI Data & Maps 2004..

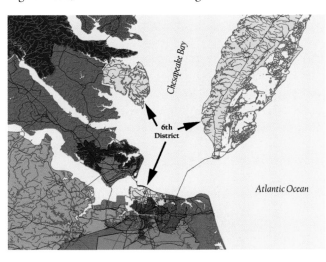

Figure 3. The 6th Virginia Senate district.

District boundary data from U.S. Census Bureau. Road and stream data from ESRI Data & Maps 2004.

House District 74, for example, is known as the "hatchet" district for obvious reasons (figure 2).

The constraints of the one-person, one-vote rule sometimes clash with the realities of geography. As a result, the 6th Senate District in Virginia is split into three parts on different ends of the Chesapeake Bay because there are not enough residents on the eastern shore to comprise one Senate district (figure 3).

Virtually every state now uses GIS in its redistricting processes to one extent or another. At the very least, the technology has made the redrawing process quicker and more efficient for legislators. One example is now an infamous gaffe concerning the home of former Speaker of the Virginia House of Delegates Richard Cranwell. In announcing the completion of the final districting plan in 2001, the Republican Party (which controlled the House of Delegates) noted that it had been necessary to place Cranwell in the same district as long-time Democratic incumbent, Chip Woodrum. It was viewed as an attempt by Republicans to force Cranwell from office. However, when the GIS maps were scrutinized closely, it was clear that the plan, as drawn, actually did not place Cranwell in Woodrum's district. Instead, the line just missed Cranwell's home by a few dozen yards.

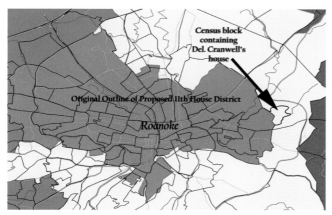

Figure 4. Gerrymandering the former Speaker of the Virginia House of Delegates. By adding a single census block to the eastern edge of this proposed district, Republicans were able to pack two senior Democrats into the same House of Delegates district.

District boundary data from U.S. Census Bureau. Road and stream data from ESRI Data & Maps 2004.

The Republicans promptly ended the session and quickly redrew the map overnight. When they returned the next day, they happily presented a revised plan that actually did situate the former speaker of the house in the same district as his senior colleague (figure 4).

Using GIS as a redistricting tool has arguably made both the process and the resulting maps more transparent to the public as well. With ready access to census data and voting results, curious amateur mapmakers can deliberate the implications of redrawn districts among themselves. Illustrating the process and providing insights into the extensive behind-the-scenes work were two objectives of our redistricting class.

Course setting

Given the spatial nature of redistricting, integrating GIS into this politics course seemed logical. Rather than teaching GIS per se, our intention was to use GIS to help students develop a deeper understanding of the course material. We divide the course time into three distinctive components: a seminar, a lab, and student project time. In the seminar, we read and discuss political science literature on voting, representation, and the redistricting process in the United States in general. Because the student projects will later focus on redistricting in Virginia exclusively, students familiarize themselves with Virginia politics by writing a research paper on the most recent redistricting process, learning while they do so which areas of the state are the most politically sensitive or controversial. For example, the western part of Virginia has been losing population, relative to the rest of the state, for some time. Accordingly, legislators from the west try to ensure that it loses as little representation in the legislature as possible. The northern and Tidewater regions of the state are intense political battlegrounds because they have been gaining population over the last two decades, and high concentrations of minority voters live in these areas (and around the state capitol, Richmond). Subsequently, redistricting in these parts of the state is always contentious because it entails the creation of new districts—that is, new zones of political influence.

The seminar is supported by a lab where students learn to use GIS software to draw and edit district maps.[7] They also learn basic numerical analysis techniques using Microsoft Excel. John Blackburn teaches this lab, guiding students through the software and data fundamentals before getting them started with their projects.

Student projects

While students research Virginia politics, study the intricacies of voting rights law, and acquire some understanding of GIS software, we charge them with studying and redrawing the state legislative district map for Virginia. Their criteria for making changes to existing district boundaries include minimizing population deviations across districts, preserving municipal boundaries, and improving minority representation.

We give the students one bit of flexibility in their redistricting process: they are free to use multimember districts. Critics blame the single-member system for many of the shortcomings of the American political system: it favors incumbents, which in turn affects voter turnout; discriminates against small parties; and subjects the redistricting process to partisan and racial manipulation (gerrymandering).[8] Political reform advocates frequently argue in favor of using multimember districts because they produce fairer and more proportional electoral results. Accordingly, drawing multimember districts improves the students' maps.

Students quickly learn that this multidimensional problem is akin to solving Rubik's cube. As soon as one district is redrawn, all neighboring districts must be checked to assure they conform to the constraints imposed by:

- the one-person, one-vote rule;
- the Voting Rights Act's requirements for minority representation;
- state constitutional requirements concerning district shape;
- the preservation of municipal boundaries; and, of course,
- the demands of incumbents.

The results have been remarkable demonstrations of creativity and political acumen. Each time we've taught the course, our students have produced highly original maps that meet or surpass the requirements posed by state and federal law. The students' district plans have all resulted in more opportunities for minority representation, the division of fewer municipal boundaries, more geographically compact and less bizarre district boundaries, and close adherence to the requirement of equal district population (figure 5).

Administering the course

Our primary course objective is for students to learn the many ways that politics, law, topography, demography, statistics, and personal relationships interact in an almost unimaginably complex process. Rather than simply reading the scholarly literature and case law on the subject, our aim is to make the learning process active and creative, with each student working to balance a multitude of considerations as they conceive and execute their own redistricting plans.

Using Chickering and Gamson's Seven Principles for Good Practice in Undergraduate Education (Chickering and Gamson 1987) as a guide, we hope to engage our students in a variety of ways:

1. Good practice encourages student–faculty contact

We note this up front as both an acknowledgment and a warning: the active, constructivist nature of the student projects and the steep GIS learning curve requires a significant amount of interaction

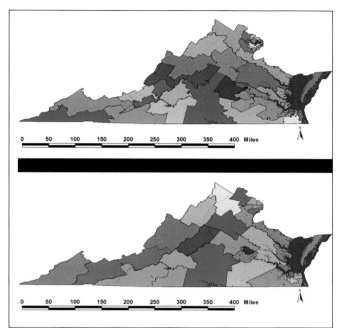

Figure 5. A Microsoft PowerPoint slide from a student presentation demonstrating the differences between the adopted districting plan for the state senate and his own, multimember solution.
Courtesy of Karl Kuersteiner and U.S. Census Bureau.

on a number of levels simultaneously. Before applying GIS technology to the problem at hand, students have to learn about Virginia politics (most students come from out of state) and redistricting law. This requires much reading and discussion time, particularly in the first weeks of the course. Throughout this course, students require ongoing guidance in both technical and disciplinary content matters.

2. Good practice encourages cooperation among students

For the first iteration of the course in 2001, students worked in three groups of three people to create redistricting plans. Students felt a strong sense of shared responsibility for learning and applying that knowledge to their team projects. In the second iteration of the course, we set up a tiny computer lab—in an electrical closet, actually—dedicated to the course and available to students 24/7. This became a kind of after-hours clubhouse for our students, who spent an enormous amount of time there, sometimes pulling all-nighters.

That year our class size was particularly small (six students) and they each drafted their own redistricting plans, but were able to learn from one another as they developed their GIS skills. Interestingly, this dedicated space created a strong collaborative spirit among the students, even though they were working on their plans individually. Our primary reason for setting up this small lab space was to bypass the typical public computer lab problems, such as security policies that prevent students from saving their work locally. Yet we found somewhat unexpectedly that this little clubhouse provided a number of significant advantages, not the least of which was peer mentoring.

3. Good practice encourages active learning

Active learning is clearly the primary strength of our approach to using GIS in this course. First we gave students a theoretical and historical context for the redistricting process. Then we equipped them with relevant quantitative, spatial, and technical skills and concepts as well as raw, unfiltered data with which they created their own redistricting plans from scratch. In this way, students came to own their learning in a way that would not have been possible if the primary learning activities had been listening, reading, and writing about this process. We note that this was a new experience for these students, one they found very enjoyable.

4. Good practice gives prompt feedback

At every stage of this course, students are required to produce some result or evidence of their learning applied to the redistricting process. For example, after the third class session, we asked students to recalculate the order in which congressional seats are apportioned by the Census Bureau. After the fourth session, they use ArcView to extract data from the Virginia state dataset and construct a discrete dataset for our local county. The technical aspects of the course are essentially self-correcting (it was obvious if something was done incorrectly), but both of us were frequently on call for students who encountered problems in the lab.

5. Good practice emphasizes time on task

The nature of their work with GIS required students to proceed carefully as they drew their legislative districts. Many reported that they repeatedly drew and redrew their maps, tweaking their districts for balance of population, race, and for the compactness of their shapes, a painstaking process that simply cannot be completed in a hurry. In the second iteration of the course, especially, when students created their plans individually and had access to a special lab at all hours of the day and night, they spent long, uninterrupted hours engaged in this work.

6. Good practice communicates high expectations

We sometimes doubted that the high volume and extreme complexity of the work at hand would "fit" appropriately within the time frame of the course. Yet we felt that these highly competitive students would rise to the challenge. We concluded that assigning the final project in the midst of our initial technical GIS training was a crucial factor in the success of this course. With the clear objective of completing an acceptable redistricting plan for Virginia by semester's end in mind, our students absorbed the necessary technical skills rapidly and with great eagerness. To have marched students through GIS training without making its purpose explicit would have been, we believe, a mistake.

7. Good practice respects diverse talents and ways of learning

This course incorporated a wide variety of learning activities: lecture (including guest lectures by outside politicians, legal experts, and issue advocates); discussion; reading; writing; statistical, spatial, and political analysis; mapmaking using GIS; and a final oral presentation to the class incorporating significant cartographic and other graphical components.

Challenges faced

Our principal challenges were technological. This discussion of Politics 295 is based on our experiences in 2001 and 2002, and the raw material of any redistricting plan is U.S. Census Bureau data. Census 2000 collected an unprecedented amount of information on race, as respondents were given the opportunity to identify up to six races in combination. As a result of this increased complexity, the Census Bureau distributed its population and race data in such a way that we were required to reprocess all of it, an undertaking that took nearly a month (of our own time, before the class even started) of head scratching, mistake making, and data processing in huge, cumbersome chunks. That was the case then, and now there are numerous versions of ready-to-map census data available. Technology advances rapidly, and by the time of the 2010 census, redistricting software may be advanced to dramatically improve the logistical aspects of a course like this. Meanwhile, we caution others that data preparation for a course can represent a significant effort.

Computer hardware and network systems become a concern with a course like this. The sheer size of the census dataset necessitates the use of local workstations for the course because performing GIS operations across a network on such large datasets degrades the performance of even a top-of-the-line server. Furthermore, we encountered performance problems with the server using census data from Virginia (a medium-sized state); operating with the datasets for larger states would be even more cumbersome. Apart from local processing advantages, students also benefit from being able to store and access their own project work and copies of the census datasets on local workstations. Implementing individual password protection or another type of security to safeguard students' files during the semester becomes necessary. Overall, we recommend that such a course be administered using the most powerful workstations available, with all data and project files located on the local hard drive.

Still, software and hardware issues notwithstanding, this experimental course was a resounding success. Our students responded enthusiastically to the opportunity to address a real-world problem. Their idealism and competitive drive pushed them to strive for an outstanding solution to this important issue that affects our individual liberties. Students thrived in the academic setting of a politics lab, which encouraged their creativity, as did regular instructor feedback and peer collaboration, both formal and informal. Our experiences demonstrated that the right combination of intriguing, cutting-edge political analysis, combined with equally cutting-edge GIS technology, provides students with a uniquely rewarding, intense learning experience.

Notes

1. The formula and process used to allocate seats among the states is straightforward. See the discussion on the Web site of the U.S. Census Bureau at www.census.gov/population/www/censusdata/apportionment/computing.html.
2. This gerrymander was, in fact, a great failure. In the ensuing legislative elections, the Federalists won the seat back. (See Griffith 1907; Hardy 1977; and Rush 1994.)
3. Baker v. Carr, 369 U.S. 186 (1962) and Reynolds v. Sims, 377 U.S. 533 (1964).
4. For congressional redistricting, the Supreme Court has maintained a strict standard of population equality, as in the case of Karcher v. Daggett (1983) where they overturned a New Jersey Congressional district map with a deviation range between the largest and smallest districts of 0.6984 percent in favor of an alternative districting plan with a smaller deviation. Karcher v. Daggett, 462 U.S. 725 (1983).
5. In Voinovich v. Quilter (1993), the Court let stand an Ohio state legislative district plan, saying that population deviations of up to 10 percent were constitutional.
6. Easley v. Cromartie, 532 U.S. 1076 (2001).
7. In the first two versions of the course, we used ArcView 3.2 supported by Autobound, a dedicated redistricting program produced by Digital Engineering Corporation (www.digitalcorp.com).
8. See, for example, Douglas 1993 and Engstromm 2001.

References

Chickering, Arthur W. and Zelda F. Gamson. 1987. Seven Principles for Good Practice in Undergraduate Education: Faculty Inventory. Racine, WI: The Johnson Foundation, Inc.

Douglas, Amy. 1993. *Real Choices/New Voices.* New York: Columbia University Press.

Engstromm, Richard. 2001. *Fair and Effective Representation?* New York: Rowman and Littlefield Publishers.

Griffith, Elmer. 1907. *The Rise and Development of the Gerrymander.* Chicago: Scott Foresman.

Hardy, Leroy C. 1977. Considering the Gerrymander. *Pepperdine Law Review* 4:243–84.

Rush, Mark E. 1994. *Does Redistricting Make a Difference?* Baltimore: The Johns Hopkins University Press.

About the authors

Mark Rush is head of the Department of Politics at Washington and Lee University. He teaches and has written extensively on constitutional law, electoral law, and comparative politics. In 1994 he authored *Does Redistricting Make a Difference?,* published by The Johns Hopkins University Press. He and his coauthor, John Blackburn, team-teach a GIS-based seminar on redistricting.

John Blackburn is head of the Instructional Technology Group at Washington and Lee University. In addition to the course he teaches with Mark Rush, he works on projects that link digital imagery to maps and is exploring Web-based mapping applications. He is a member of the GIS tutorials design team for the National Institute for Technology and Liberal Education (NITLE).

Many students today choose interdisciplinary majors, sensing the problems of the world are too complex to study from a single vantage point. To understand global warming, teen pregnancy, gang violence, or disease transmission, one must be trained in multiple fields; work from several disciplinary perspectives; find, handle, and coordinate data from disparate sources; and communicate findings clearly to a broad audience. But how can anyone learn so much in four years of undergraduate education, given the fearsome pyramid of prerequisites? How can we teach our interdisciplinary majors to weave multiple perspectives together to better understand—and work to address—the problems of the world?

GIS may serve as a unifying technology in interdisciplinary studies, allowing us to combine data from multiple sources, analyze that data both qualitatively and quantitatively, and communicate our findings to a general audience in both graphical and statistical format. While 93 percent of our liberal arts colleges and universities lack formal geography departments (Bjelland 2004), the diverse methodologies of the geographer may still offer "a path to the integration to which a liberal education aspires" (Bjelland 2004, 331). The purpose of this chapter is to persuade others teaching in interdisciplinary fields at liberal arts institutions that geographic information systems may be a worthy addition to our curricula. In this chapter I will describe my own experience integrating GIS into an already rich urban studies curriculum.

Chapter 10

Urban studies: Assessing neighborhood change with GIS

Christine Drennon

Approaches to interdisciplinary studies

Interdisciplinary studies (IDS) differ fundamentally from classical disciplinary studies in their history, ideology, and focus. In fact, interdisciplinary studies in liberal arts colleges may present an oxymoron. A traditional "liberal education" centers around the classical humanist traditions passed down from the writers of ancient Greece and Rome, and has traditionally been concerned with ideas in the abstract, with the conservation of essential truths, and with the development of the intellect (Miller 1988, p. 183). Interdisciplinary studies, on the other hand, "may be defined as a process of answering a question, solving a problem, or addressing a topic that is too broad or complex to be dealt with adequately by a single discipline or profession" (Klein and Newell 1997, p. 393). IDS are concerned more with experimentation and problem solving than with abstract knowledge, and have found their authority in contemporary needs, rather than in the ideas of classical authors. The issue is prioritized over the foundation—a worthwhile approach to the complexities of the early twenty-first century, but a challenging one.

The difficulties of interdisciplinary research, teaching, and communication are not new. For example, understanding the science behind public health issues and sharing the information in a meaningful way requires an interdisciplinary approach. In 1854 Dr. John Snow argued persuasively that cholera was transmitted through water. Between 1849 and 1854, more than 14,000 people died of cholera in London. The 1854 outbreak was equally virulent; 324 people died in the first four days of the outbreak (Snow et al. 1936). This was a true urban crisis.

As he worked to understand the epidemic, Snow gathered geologic information (substrate), demographic data (who had died, where they lived, how old they were, and what they did for a living), and medical data (including his knowledge that a gastrointestinal disease such as cholera was probably ingested rather than inhaled). He mapped occurrences, looking for spatial patterns that had heretofore gone undetected. He was acting not only as a medical doctor—he was also drawing on knowledge of geology, economics, sociology, and geography to solve a contemporary crisis. The sheer amount of data required, in combination with his own radical rejection of vapor theory and contemporary assumptions about disease transmission, allowed him to arrive at a hypothesis and communicate that hypothesis visually to a more general, nonscientific audience.

John Snow's analysis of cholera transmission in the Golden Square neighborhood in London was only one of three contemporaneous studies (Melnick 2002). The Cholera Inquiry Committee of the vestry of the Parish of St. James, Westminster, conducted its own survey, as did England's General Board of Health. Two of the three studies—Snow's and the Cholera Inquiry Committee's—reached the conclusion that cholera was transmitted through the water; the third, the General Board of Health, reached a different conclusion: that "the deadly nocturnal vapors emanating from the Thames River caused the epidemic" (Melnick 2002 p.2). All three investigators constructed maps to analyze the situation. All used dots to represent deaths from cholera, yet they reached different conclusions because the manner in which they used the maps differed; the Board of Health looked at the concentration of cholera deaths in the neighborhood while Snow looked at both the presence and absence of deaths in the same neighborhood, analyzing the spatial distribution of occurrences, and thus reaching a very different conclusion. Although his GIS was rudimentary, the power of his analysis was far-reaching (figure 1).

Figure 1. Snow's cholera map analyzed the presence and absence of deaths in the same neighborhood of London.

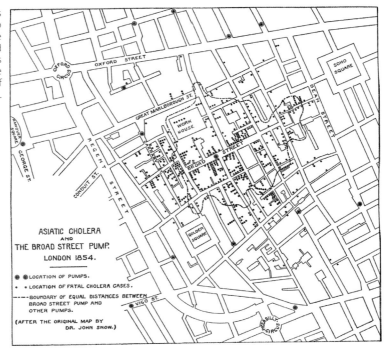

We still have health crises in cities, as well as crises of the environment, safety, fiscal matters, family, education, and security. To address any one of these, one must draw from a breadth of knowledge and disciplines, compile data from numerous sources, and analyze how it converges spatially. A geographical information system, even one as primitive as that of John Snow's, allows us to tackle such crises. But how do we teach students in our interdisciplinary programs to be thorough researchers and effective communicators, as Dr. Snow was? And how do we teach those techniques in a four-year program, on top of all of the requisites and prerequisites that our students already face? How do we introduce yet another subject into an already full curriculum—especially another subject that will introduce unfamiliar concepts of geography and mapping to students who had not set out to study those subjects?

By mirroring the ideology that underpins any interdisciplinary major, GIS can be taught in a problem-based manner, centering on the problem rather than the technology or the discipline. I present GIS as a research tool comparable to those used in other social science fields (e.g., statistical analysis with SPSS software, or survey methodology). I do not approach GIS from the realm of cartography and thereby avoid prerequisites in the field of geography. I rely only on prerequisites shared by the social sciences such as statistics and research design, already part of our IDS curriculum.

Problem-based instruction

Boud and Feletti (1991) detail the process of problem-based instruction (see also Shepherd and Cosgriff 1998):

- Students are presented with a problem; students then work in groups to organize their ideas and previous knowledge and to define the broad nature of the problem.
- Students begin to articulate questions that reflect aspects of the problem they do not understand. Students define what they know and what they don't know.
- Students rank the issues or questions articulated and divide the labor among themselves to pursue the answers. They define the resources necessary to research the issues. In the case detailed here, the resources included knowledge of the history of the issue, socioeconomic data, and GIS skills to compile and analyze the data.
- Upon reconvening, students present and summarize their findings and, as a group, decide how to proceed.

In this way, students are presented with a problem and then engage in a "self-directed, reiterative, reflective learning process" (Shepherd and Cosgriff 1998, 350; see also Barrows and Tamblyn 1980). The iteration ensures that new skills are mastered, while the self-direction and room for reflection create a learning experience in which authority is decentralized—if not reversed. In this way, GIS can be taught to students already over-burdened with a full curriculum. I do not set out to teach my students a complete course in GIS, something beyond the scale and scope of our IDS curriculum. Rather, students frame their learning themselves by first identifying the questions they must answer to solve their problem. My GIS instruction then supports those inquiries.

The project described here was undertaken for Habitat for Humanity (HfH), in San Antonio, Texas. Habitat for Humanity has been active in San Antonio for more than thirty years, and has built more than three hundred houses. Despite its longevity, the organization has never had the means to conduct an assessment of its impact on the city or on the families it has helped.

This assessment was the problem posed to students on the first day of class, a semester-long project that HfH commissioned within the context of my GIS course. While a majority of students in this class were urban studies majors, others came from the departments of political science, economics, and sociology. Thus not only were there true interdisciplinary majors (the urban studies majors are required to take classes from several different disciplines), but the class itself is interdisciplinary. It can be challenging to teach students who do not share a common perspective or knowledge of any one particular discipline. However, here it worked as an advantage because we required a wide variety of disciplines, perspectives, and methods to thoroughly understand our real-world problem. The course itself was relatively unstructured. As we worked through particular issues, students recognized their need to answer particular questions. Overall we followed a linear sequence through stages of framing our questions and analyzing and communicating our answers.

Step 1: Articulate questions

On the first day of class, I told students the overarching objective of the course: they had fourteen weeks to complete an assessment of (1) the impacts of HfH on the city of San Antonio, and (2) the impacts of HfH on the families served. At the end of the semester, they would present their findings to the local board of directors of Habitat for Humanity. After much consternation and confusion, students began to ask a series of questions that served as a guide for the beginning of the project.

1. Where are the HfH houses?
2. What are the demographics of the HfH communities?
3. What are the demographics of the larger communities and neighborhoods in which the houses are located?
4. What variables should be tested to determine if there are differences between communities in which HfH has been active and where it has not?

Students quickly realized they lacked the mapping skills necessary to accomplish such a project. It was at that point—very early in the process—that I introduced GIS.

The course objective, to assess the impacts of HfH, dictated how and when I would teach particular GIS software skills. Because the course objective was inherently geographical, integrating the GIS technology was straightforward and timed to allow students to use their new skills to solve problems. In addition, because the course objective (and not the technology) served as the main focal point of the class, students frequently envisioned their mapping needs before learning the capabilities of GIS.

When teaching GIS to students who have never encountered it before, I divide it into four of its unique components:

- Working with maps as layers
- Linking spatial locations with attributes
- Using and appreciating topology
- Communicating findings graphically

I introduce various GIS techniques according to this logic. Within each they demonstrate their mastery of the GIS skills. Evidence of mastery may be answering a simple question requiring some data manipulation or producing a map.

The first task that the class set for themselves was to know where each HfH house was. In addition, they agreed they needed to know for whom the houses had been built (the characteristics of the new home-owning communities), and finally, they asked in what kinds of neighborhoods is HfH active? All of these socioeconomic variables are readily available from the U.S. Census Bureau, and can be downloaded into a GIS. Therefore, our first task was to produce a map we called "status quo."

The status quo map was multilayered, therefore its production exposed students to this first essential GIS feature: maps as layers. As shown in table 1, several elementary GIS skills are introduced at this stage, including downloading, displaying, and navigating the layers.

Early in the project, students were expected to complete their own work and hand in their own assignments. However, our environment was a problem-based learning classroom that encouraged group work and shared information. Students who had difficulties felt free to ask one another, although ultimately they were each responsible for knowing the processes themselves. By the time

Table 1. Learning GIS through skills and tasks

GIS component	Individual skill being introduced	Tasks assigned to learn that skill
Using maps as layers	Downloading spatial data	Downloading TIGER/Line files from the Census Bureau
	Importing data	Bringing boundary line files into ArcGIS
	Navigating a map	Zooming, panning
	Looking at attributes	Using the identifier button
	Symbolizing data	Symbolizing features by categorical attributes
Exploring the link between maps and the attribute table	Investigating the link	Selecting records
	Querying data	Selecting by attributes
	Performing joins and relates	Joining layers with data tables
	Geocoding	Geocoding tables and linking to TIGER/Line files
Appreciating the power of topology	Using topology	Querying layers; location queries
	Using topology	Combining attribute and location queries
	Spatial relations and topology	Selecting by location; buffering, overlaying
Communicating findings graphically	Basic mapmaking skills	Setting projections, laying out

these assignments were completed, everyone was familiar with the skills in table 1, and some were even confident, at which point it was time to begin the analysis.

Step 2: Analyze data

Producing the status quo map accomplished three things: it confirmed the students were beginning to master introductory GIS software skills, it informed the next stage in the project, and it revealed—what to the students were—some unexpected and surprising spatial relationships (figure 2). To facilitate these revelations, Stage 2 began with a brainstorming session. Our original question was "What impact has HfH had on the neighborhoods of San Antonio?" To answer this question, students had to decide how they could determine and measure if HfH communities differed from the surrounding neighborhoods or not. They chose four indicators: education (Do children in all HfH neighborhoods attend school? Do standardized test scores in their schools differ from neighboring schools' scores?); crime (Do crime rates in HfH communities differ from neighboring communities' rates?); investment (Do investments made by HfH homeowners in their homes and neighborhoods differ from their neighbors' investments?); and community participation (Does the participation by HfH homeowners in their local communities differ from the level of participation by non-HfH homeowners in the same communities)? Figure 2 shows examples of the maps the students produced.

To address these questions, students broke into four groups to work on these different indicators. During this time, they refined many of the GIS skills learned in the first part of the course. They had to identify data, import it into the GIS, analyze it, and explore relationships between those variables. Learning the second set of GIS capabilities (see table 1), students became familiar with skills such as selection by attribute. Exploiting the GIS link between the spatial and the numeric, they were able to filter the data to create maps that revealed both qualitative and quantitative relationships. As they became more familiar with their own data and the possible relationships between layers, they moved to the third distinctive component of GIS: its use of topology.

Topology, which deals with geometric properties of objects that remain invariable when the object is bent or stretched (such as in a map projection), is a complex branch of mathematics, and its study is beyond the purview of my urban studies class. An appreciation for it, though, and an understanding of what it allows us to do, is critical to the success of this and every GIS project. To guide them into an appreciation of topology, I require them to experiment with two quantitative spatial analysis techniques: buffering and spatial overlay. With gentle prodding, I encouraged the different groups to form questions that would require the use of these two analysis techniques. These techniques allow students to realize that layers are "aware" of one another—that they "know" the spatial relationship between themselves. This appreciation helps students form spatial relationship questions in the future.

The education group, for example, was charged with the task of determining if the schools attended by HfH children differ from schools that do not serve the HfH communities. To answer that question, they needed to determine the neighborhoods that fed into the various city schools. This was done by first geocoding the addresses of the schools in the county and creating a point-data layer of schools, and then buffering the HfH neighborhoods to determine their proximity

High School Drop-Out Rate in San Antonio, TX

| • | HfH_houses | % drop_out | | 0.043078 - 0.103704 | | 0.189724 - 0.318996 |
| | | | 0.000000 - 0.043077 | 0.103705 - 0.189723 | | 0.318997 - 0.572277 |

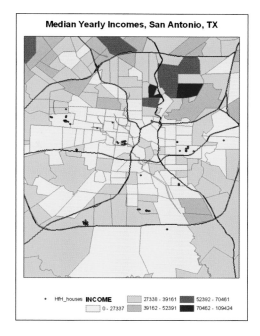

Median Yearly Incomes, San Antonio, TX

| • | HfH_houses | INCOME | | 27338 - 39161 | | 52392 - 70461 |
| | | | 0 - 27337 | 39162 - 52391 | | 70462 - 109424 |

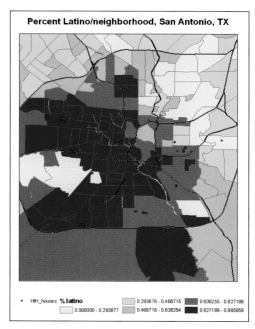

Percent Latino/neighborhood, San Antonio, TX

| • | HfH_houses | % latino | | 0.293678 - 0.466715 | | 0.638255 - 0.827188 |
| | | | 0.000000 - 0.293677 | 0.466716 - 0.638254 | | 0.827189 - 0.985858 |

Figure 2. Examples of the maps that students used to explore the spatial relationships between locations of Habitat for Humanity houses and social and demographic variables.

Data from ESRI BIS.

to various schools using a constant distance (figure 3). Buffering is an inexact approximation of a school-catchment area but suffices in the cases studied here because of the close proximity of the major HfH neighborhoods to local elementary schools. Once students determined which schools the HfH children attended, they could make comparisons with the performance of children at all San Antonio schools.

In addition to buffering, students were directed to explore their datasets by overlaying them in different combinations. The group charged with the task of investigating investment in and management of the home (Do HfH homeowners invest differently in their property than non-HfH homeowners?) had to determine how they would measure investment (they chose the purchase of a central air conditioning unit as an indicator) and how they would measure management (they

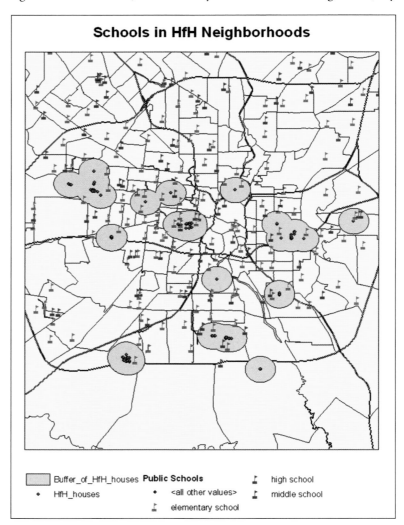

Schools in HfH Neighborhoods

Figure 3. Public schools located in the vacinity of Habitat for Humanity housing

ESRI Business Information Solutions. Data from author and ESRI Data & Maps 2004.

Buffer_of_HfH_houses	**Public Schools**	⌐	high school
• HfH_houses	• <all other values>	⌐	middle school
	⌐ elementary school		

chose city-code-compliance violations). They edited attribute tables to include this information and combined and clipped layers to illustrate where such actions had taken place.

Finally, the group investigating crime statistics used both qualitative and quantitative analyses to determine if personal and property crime rates differed by neighborhood (figure 4). Once they had visually examined and compared the patterns to determine whether a relationship between the two (a "qualitative" inspection of the data) existed, they performed a traditional, parametric statistical assessment (t tests) to see if crime rates were the same within and outside the neighborhoods. This required strict and careful comparisons of the number of crimes committed and the borders of all neighborhoods. In addition, they also compared the types of crime committed. In this way, they used both their new GIS skills and their statistical skills to analyze the data.

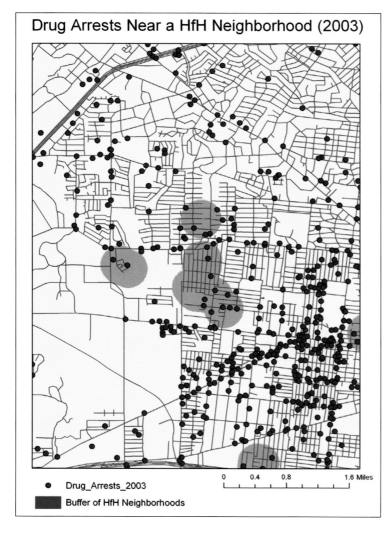

Drug Arrests Near a HfH Neighborhood (2003)

0   0.4   0.8   1.6 Miles

● Drug_Arrests_2003

▬ Buffer of HfH Neighborhoods

Figure 4. Crime patterns surrounding Habitat for Humanity housing.

Data from author and ESRI Data & Maps 2004

Step 3: Communicate findings

The final topic introduced in this GIS class was mapmaking. Introducing mapmaking at the end of the course, as opposed to the beginning (which is how it has traditionally been presented in geography departments where cartography is a prerequisite to GIS classes), may be one of the major differences between this project-based course and a more traditionally presented class. While basic mapmaking techniques (for example, what should be included on a map, color choices, scale of presentation) are presented at this time, it is difficult to reconcile the intricacies of mapmaking with the technology of GIS in one fifteen-week semester. New books (e.g., see Krygier and Wood 2005) are addressing this lacuna in the cartography and GIS literature.

While completing their data collection and analyses, students created many illuminating maps that furthered their own insights. They also learned mapmaking skills as the project progressed. Though the groups worked fairly independently from one another, they reported their findings to the class weekly, a useful preparation for communicating their work to the sponsors.

Finally, in the third and final month of the project, students presented their findings and recommendations to an expert panel consisting of the board of directors for HfH, staff at HfH, and members of the university community. Though the map was the primary form of communication, students also prepared a written report to explain many of the mapped features as well as the process by which the map was produced. The ability to present numerical tabular data in a nonnumeric, spatial format enabled students to communicate their findings to a general audience. Moreover, presenting data as images, rather than as tables and graphs, made their analysis accessible to listeners with highly diverse backgrounds.

Reflections on teaching and learning

This course challenged students on many levels, though all challenges were both realistic and surmountable. Some students felt uncomfortable in the role of framing assessment questions requested by San Antonio Habitat for Humanity. Others felt anxious as they learned to use new technology. Still others were forced to collaborate across majors, sharing their varied background knowledge. In their analyses and presentation they were required to bring together many disparate pieces into a coherent description of HfH's impact on local communities.

Though these students rose to every challenge, I perceive problem-based learning to be most successful with advanced and motivated students. My students were predominantly juniors and seniors, and I have found that younger students often lack the content knowledge and experience to ask the relevant, guiding questions that, in turn, motivate them to learn the somewhat complex GIS technology.

In this course, I evaluate students according to several criteria that differ significantly from more traditional classes. Because the main focus of this class is on the project, and GIS skills are acquired as needed, students are assessed in two ways: on their participation in the learning process and on their successful mastery of key GIS skills. I base my assessment on their commitment to participate in group work (which is also assessed by their peers), occasional intermediate deadlines (when groups present their work to one another), and their level of commitment to the final project. Because this is a problem-based class in which students determine many of the details of the study, they encounter authentic hurdles well known to every researcher. Sometimes ideas don't bear fruit. Data may not be available, or may not be accurate. Research questions may simply not be well developed. However, if they attend to the process of learning GIS in this problem-based manner, then all students are rewarded for their efforts, regardless of the obstacles they encounter. In addition, all students take a final exam that tests their GIS skills independently. This final exam question is typically a problem inspired by a newspaper story. I asked this class where the next HUD (Housing and Urban Development) housing development should be placed in our city and gave them a certain set of priorities and criteria. The final exam takes three hours to complete, during which students use maps and GIS processes to answer the question and identify a specific census tract.

Though the learning of GIS was secondary to the overall objective of this course, GIS made it possible for the students to reach the principle goal: to perform and deliver an assessment of HfH's impact on San Antonio. GIS allowed the students to collect and combine data from scattered, spatially inconsistent sources (including the U.S. Census Bureau, the State Board of Education, the County Appraisal Office, and a precinct's database), to examine and measure the relationships between the data, and to communicate their findings to a general and diverse audience.

Problems encountered in an interdisciplinary studies class mirror problems encountered in the world: they are multifaceted, they involve data gathered from disparate and inconsistent sources, and they may require complex spatial statistical analysis. GIS is an ideal unifying tool and method for meeting these challenges. Despite the additional complexities that GIS introduces, it is a valuable addition to an urban studies curriculum, especially in a project-based learning environment emphasizing the problem over the technology.

References

Barrows, H. and Tamblyn, R. 1980. *Problem-based learning: An approach to medical education.* NewYork: Springer Publishing Company.

Bjelland, M. 2004. A place for geography in the liberal arts college? *Professional geographer* 56(3): 326–36.

Boud, D. and Feletti, G. 1991. *The challenge of problem-based learning.* New York: St. Martin's Press.

Klein, J. and W. Newell. 1997. Advancing interdisciplinary studies. In J. Gaff and J. Ratcliff, eds. *Handbook of the undergraduate curriculum: a comprehensive guide to purposes, structures, practices, and change.* San Francisco: Jossey-Bass Publishers.

Krygier, J. and D. Wood. 2005. *Making maps: a visual guide to map design for GIS.* New York: Guilford Press.

Melnick, A. 2002. Introduction to geographic information systems in public health. Gaithersburg, Maryland: Aspen Publishers.

Miller, G. 1988. The meaning of general education: the emergence of a curriculum paradigm. New York: Teacher's College Press, Columbia University.

Shepherd, A. and Cosgriff, B. 1998. Problem-based learning: a bridge between planning education and planning practice, *Journal of planning education and research,* 17, pp. 348–57.

Snow, J., W. H. Frost, and B. W. Richardson. 1936. Snow on cholera. London: Oxford University Press.

About the author

Christine Drennon is an assistant professor in the Urban Studies Program at Trinity University in San Antonio, Texas. Her academic background is in geography, but she did not use GIS to a great extent until she began to teach at Trinity. She is committed to teaching GIS as a tool for her students' use in exploring and analyzing urban phenomena. Her research focuses on social reproduction and schooling in urban environments.

The Collaboratory for GIS and Mediterranean Archaeology (CGMA) is an interdisciplinary, inter-institutional endeavor that involves faculty at four undergraduate institutions: DePauw University in Greencastle, Indiana; The College of Wooster in Wooster, Ohio; Rhodes College in Memphis, Tennessee; and Millsaps College in Jackson, Mississippi. As one part of CGMA, DePauw has offered a seminar on Archaeology and GIS since 2003 as a collaborative effort across the four institutions. This seminar covers the methods, theories, and practice of field or lab research with library research in archaeological survey. Students are deeply immersed in the application of GIS to archaeology and apply that learning through a multistage practicum involving GIS-based survey work. CGMA students are also integrally involved in creating MAGIS (Mediterranean Archaeology GIS), our Internet-based GIS of metadata—data about data—from archaeological survey projects in the Mediterranean basin. For scholars and students worldwide, MAGIS provides an inventory of exactly what work has been done where, when, and how, allowing scholars worldwide to combine, and perhaps compare, survey data across the Mediterranean.

Classical archaeology: Building a GIS of the ancient Mediterranean

Pedar W. Foss and Rebecca K. Schindler

Many archaeologists use GIS as a critical research tool, but it is seldom a focus of comprehensive instruction for undergraduate classical archaeology students.[1] In 2000 when we originally conceived of Collaboratory for GIS and Mediterranean Archaeology (CGMA), we did not envision an instructional component, but over the years the concept evolved so that teaching the theory and practice of GIS became a defining characteristic of the project. As archaeologists, we had previously learned GIS concepts and acquired relevant software skills as needed to apply the software mapping tool effectively to our specific research projects. We now recognize additional benefits of collaborative, inter-institutional teaching: our individual areas of expertise with GIS can be combined to generate the substantial, collective knowledge that fuels our instruction.

Intercampus collaborative (ICC) teaching can be an effective approach for certain academic subjects at small colleges and universities. Individually, our small departments do not have the luxury of maintaining enough faculty to offer either the breadth or depth that larger universities can. The Sunoikisis Project, a series of ICC courses in Classics, developed and successfully operated for several years by the Associated Colleges of the South (ACS), responded to these challenges by working collaboratively across multiple small college campuses.[2] We modeled CGMA after that program. The design of the Collaboratory pivoted around a multivalent intersection between peer institutions, teaching, and research, and the kindred disciplines of classical archaeology and anthropology. This ideally suited our roles at colleges where teaching is privileged, research is increasingly valued, and opportunities for students to apply their learning to real-world problems are prized.

As it now stands, CGMA has two parallel and equally important objectives—creating wider opportunities for undergraduate students to engage in the full process of archaeological research and creating a spatially searchable database of archaeological survey projects in the Mediterranean (MAGIS, or Mediterranean Archaeology GIS). Students from various disciplines (e.g., classical studies, anthropology, computer science, and geology) are engaged in different aspects of the project, and this diversity of backgrounds enables the students to participate in all stages of forming, constructing, testing, and using MAGIS.

Students have the opportunity to work on the CGMA project in three ways: the ICC seminar in the fall semester, spring-term work-study grants, and summer research internships. Through CGMA, we also complement the ACS Archaeology program, which offers an online archaeology course in the spring term and takes students to Turkey and Mexico for fieldwork in the summer.[3] Because MAGIS is such an integral component of the CGMA project, we will first describe the research problem that MAGIS addresses, and then how we involve students with all aspects of MAGIS and CGMA.

CGMA's approach to archaeological research in the Mediterranean

Over the past forty years, archaeology has shifted its emphasis from single-site investigations to regional studies, allowing archaeologists and historians to ask not just synchronic (single-period) questions, but broadly based diachronic (multiperiod) questions as well. Technological advances in data collection, storage, and analysis supported this change. One of the most important advances has been archaeologists' use of GIS, which permits us to present our analyses of spatial data through maps. (All properly collected archaeological objects have a spatial context that is crucial to their interpretation.)

Unfortunately, most of these datasets remain isolated from each other, trapped in localized and often proprietary databases. Individual scholars have pursued pan-Mediterranean studies of archaeological data, but their compilations are in analog form (often a catalog), making it difficult for other researchers to experiment directly with the same data. Alcock and Cherry (2004) recently examined the comparability of survey data collected in regional studies, and their important contribution includes a long list of Internet resources for survey projects. However, their work and other similar projects fall short of what is needed: a dynamic resource that is constantly updated with data from new studies. A comprehensive resource could assist archaeologists, anthropologists, and historians to investigate questions of social and political development, cultural interaction and diffusion, and economic exchange. Through a spatial interface, users could query the data and generate geographic visualizations from vast stores of site-specific data.

The technology exists to make such a large-scale project feasible and productive: secure, long-term, and high-capacity storage; powerful relational databases; GIS; and the Internet. Yet the conceptual and categorical conflicts inherent in the data are formidable. Practically every team of researchers has constructed different definitions for their data and different procedures for collecting and studying that data. Accordingly, the primary rubric offered by the Alcock and Cherry volume is side-by-side rather than head-to-head comparison. In other words, the data itself cannot be compared—only the general patterns indicated by the data. This diversity of data and process poses a major hurdle for doing unified searching and making integrated comparisons.

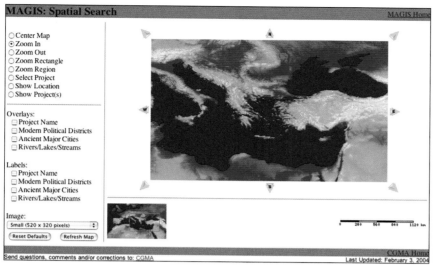

Figure 1. The area of interest for MAGIS, as seen through its Spatial Search page, beta version. The full extent is a rectangle extending from 20° N latitude, 15° W longitude in the southwest to 60° N latitude, 50° E longitude in the northeast. This area includes fifty-three countries, and covers the "extended Mediterranean" region (those regions connected to or significantly influenced by ancient Mediterranean cultures up to about AD 700).

Of course, before we can conceive of comparing survey data from the rich and diverse collection of Mediterranean area research, the data must first be collected and categorized. As a stepping stone, CGMA has been constructing MAGIS, a basic Internet GIS that serves as a geodatabase of metadata for Mediterranean survey projects.[4] The objective of MAGIS is to define and monitor the scope and scale of the worldwide body of research, as it grows. To build MAGIS, CGMA scholars and students first select archaeological survey projects and extract information from these surveys about what data was collected and how the data was stored. This information is then organized, stored, and linked to a detailed topographic map of the greater Mediterranean region (figure 1).[5] Scholars worldwide will be able to query survey data either spatially, by choosing a location, or through specific fields, by searching the data (figure 2). Our user interface, therefore, communicates both with the database and with the GIS software, and returns search results represented either in a list or a map, as the user prefers.[6] MAGIS will contain comprehensive and current bibliographies for each survey project, and therefore should become a first stop for anyone who wants to research archaeological surveys in the greater Mediterranean within the project limits.

The CGMA fall seminar and practicum

MAGIS and CGMA are profoundly intertwined, and students are integrally involved in the ongoing development of MAGIS, particularly during the ICC fall-semester seminar course on survey archaeology and GIS, which involves all four college campuses. This seminar also incorporates a practicum for survey research that is conducted by the individual instructors on the four respective campuses.

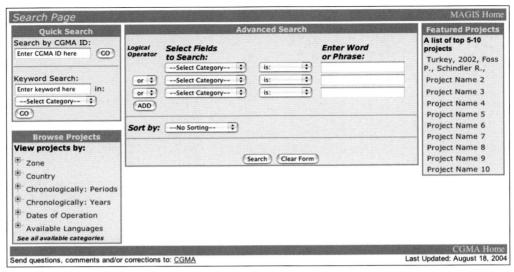

Figure 2. The database search page for MAGIS, beta version. Users have several search options: a simple keyword search, an ID number search (if the user has a favorite project, for which one knows the ID number, and wants to access it quickly), and an advanced Boolean search through multiple fields. Users may also browse through projects (lower left) by clicking through nested lists of geographic zones, modern countries, chronologies, and languages in which reports have been published. A list of most-requested projects is available on the right-hand side.

The seminar

We designed our semester-long ICC Seminar in Archaeology and GIS to introduce advanced undergraduates to methods, theories, and practice in primary (field or lab) and secondary (library) research in archaeological survey. Even though we have four institutions spread across several states participating in this ICC, the group size remains small (three to four students per school per year) and we create ample time to get to know one another through online chatting and at least one face-to-face meeting. The seminar meets synchronously online twice a week using the ACS Course Delivery System (CDS),[7] and the role of the lead instructor for the semester alternates among the four participating colleges. The lead instructor has primary responsibility for coordinating the syllabus and consults with other faculty members to develop lectures, exercises, and discussion on the history, methods, and theories of archaeological survey research. Moreover, each faculty member individually leads the students from their particular campus through a multistage practicum involving GIS-based survey work.

Throughout, students conduct individual projects that contribute to the construction of MAGIS. These projects are designed so that computer science students can work on databases, GIS, or Internet delivery programming under the supervision of the project programmer, and students in classical studies, anthropology, or geography can work on the collection, organization, and assessment of metadata from published survey projects in specific regions of the greater Mediterranean. For example, in 2003–2004, students tracked down surveys conducted in Cyprus, Greece, FYROM (Former Yugoslav Republic of Macedonia), Albania, Serbia-Montenegro, Austria, Germany, Italy, Tunisia, Jordan, and Israel.

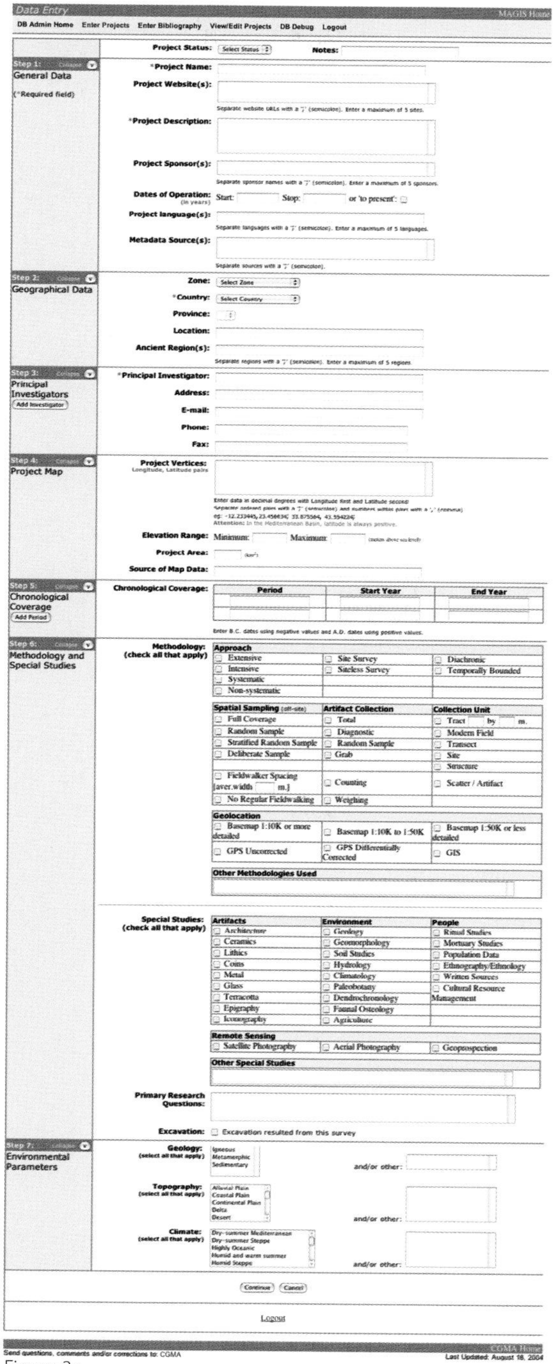

Figure 3a

A searchable database is only as good as the data within it. As part of their involvement with CGMA, students spend several hours each week outside of class time systematically wading through archaeological bibliographic databases, indices, and the Internet for references to field survey projects. They investigate the references, order necessary publications through interlibrary loan, and then comb through them in search of metadata that meets the criteria of the MAGIS categories. Once faculty have reviewed and approved these findings, students enter the data through a Web interface (figure 3). Researchers can enter data about their own sites through a data entry form on the Web. We recognize, however, that researchers and faculty from every discipline are notoriously bad at submitting and updating metadata. Thus the thorough, conscientious, and hard work of our undergraduate students will probably continue to be the primary method of data collection.

In the seminar we balance lectures and discussions, library-based research, GIS skills acquisition, and field work (the practicum). Participation in the development of MAGIS teaches students fundamental bibliographic research skills applied to archaeology, skills that are essential for graduate school but rarely taught at the undergraduate level. Library research is

Figure 3. The data entry and bibliography pages for MAGIS, beta version. Each step listed on the left-hand side of figure 3a is collapsible, making it easy to view and enter only the data one wishes to enter. Students are asked to first enter the data on a paper version of this page, so that there is a backup in case the Internet connection is lost during the entry process. Bibliographical references are entered in a separate, related database such that records can be associated with more than one project (figure 3b).

Chapter 11 Classical archaeology
Building a GIS of the ancient Mediterranean

DB Admin Home   Enter Projects   Enter Bibliography   View/Edit Projects   DB Debug   Logout

Search Options:		Search by:			
		☐ ALL			
Author's Last Name:	ALL ⊕	☐ Author	☐ Date	☐ Title	
Keyword Search:	enter keyword here	☐ Editor	☐ Edited Volume	☐ Journal	
(Submit Search)		☐ Series	☐ Place	☐ Publisher	

Displaying Results from: **0** to **0** . Total number of results: **0**

(<<) (>>) (Could not find any entries matching your criteria ⊕) (Add to Table) (Details)

☐	Author(s)*	Date	Title	Editor(s)*	Edited Volume	Journal	Series	Vol	Place	Publisher	Pages	Notes	Status
☐													Select Status ⊕
☐													Select Status ⊕

(Delete Selected) (Add blank row)   (Submit Bibliography Data) (Cancel)

*Notes:

- Please enter names starting with Last name, followed by a comma and then the First name, title , etc.
- Separate individual names with a semicolon. Example (Last1, First1; Last2, First2; Last3, First3; , etc...)
- For bibliographical references with no authors or editors listed, please enter **AA.VV** (for "autori varii") in the authors' field.

Logout

Figure 3b

more laborious than glorious, but we consider this practice as important as the more popular field experience gained on digs in the Mediterranean. Furthermore, this approach also exposes students to the importance of learning the various modern languages in which the research is published.

One of our most important sessions is the face-to-face meeting that happens midway through the semester, when all students and faculty gather at the campus of that year's lead instructor. This meeting is an integral part of the collaborative instructional mission for several reasons. First, during that time we provide students with an introduction to ArcGIS software, so that each campus team can use the software to carry out their distinct practicum.[8] There is also the opportunity to train students in data collection with handheld GPS (global positioning system) units and/or a laser transit (figure 4) that they can then use during the ArcGIS training.

Second, we have an open session, with students, faculty, and visiting CGMA board members, reporting on the current progress of, and problems facing, the CGMA project (figure 5). During this session the scholars model discussion and debate as they hash out the issues at stake, and students are encouraged to participate actively. The purpose is not only to solve the immediate problems or at least come up with promising approaches, but also to involve students in the intellectual engineering process of making a large collaborative project work. No scholar operates an archaeological field

Figure 4. CGMA students learning how to use the laser transit at The College of Wooster in October 2005. Students collected points on Wooster's campus with the transit (as well as with handheld GPS units) and then plotted those points during the ArcGIS sessions. Although most of their projects do not require the use of a laser transit, this training is useful for those students who plan to work on archaeological projects.
Photo by Rebecca Schindler. Used by permission.

Part 2   GIS case studies in the curriculum

Figure 5. CGMA faculty and students collaborating in a breakout session at the midterm meeting of all campuses at The College of Wooster in October 2005. Typically at these meetings, we pair a faculty member with a subset of the students to work on a specific theoretical, methodological, or practical problem facing the project, such as how to design the variable of "chronology" such that calendar years and named periods can both be included as options. These sessions have been invaluable; on multiple occasions important new ideas or solutions for problems have come directly from the students.

Photo by Rebecca Schindler. Used by permission.

project alone, and so working together fruitfully (even if everyone is not in complete agreement) is a necessity if the project is to be productive and successful. Students rarely have the chance to observe this behind-the-scenes process. We expect them to apply these skills as they participate in the practicum on their campus, conducting a local survey project of their own design.

The practicum

At that point in the semester when students have been exposed to the fundamentals of both survey research and GIS software, and our expectations for their collaborations have been made clear, students at each of the four campuses are responsible for designing, implementing, and analyzing a survey of part of their campus or town. The projects themselves have varied widely. From 2003 to 2004, students at the College of Wooster surveyed two cemeteries, and Millsaps students carried out two projects: one Cultural Resource Management GIS of their campus and a survey of a Jewish cemetery in Jackson, Mississippi. Rhodes students completed an efficiency analysis of the distribution of faculty offices to the classrooms in which they teach, and tracked the performance venues of Elvis Presley in Memphis, Tennessee. DePauw University students surveyed a local cemetery (figure 6), and mapped the distribution and history of churches in Greencastle, Indiana (figure 7). The findings of these surveys are not as important as the methods; our primary goal is to have the students apply professionally and collaboratively the techniques they have learned.

By the end of the term, students merged their maps and databases in ArcGIS to produce some basic maps. In figure 7, 10-by-10-meter sample squares are visible as clusters on the aerial photograph. One of the initial questions being asked about this cemetery was the history of its spatial development. At this stage it became clear to the students that a 20-percent random sample was insufficient to address that question. This was a good lesson in how initial results sometimes require additional research to capture legible and meaningful patterns.

As facility with ArcGIS increased on the member campuses, students began to download and georeference aerial photos, topographic maps, and other materials common to GIS for archaeological surveys. They collected the geographic coordinates using GPS. A problem common to all the practica has been an inadequate amount of time during one term to collect enough information to map out and analyze patterns. Extending a particular practicum over multiple years might address this issue.

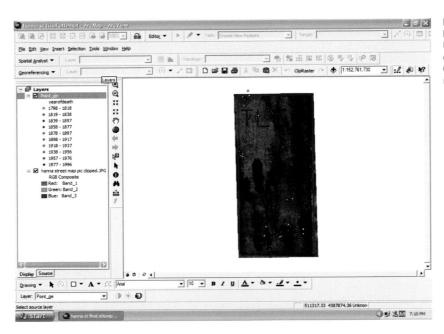

Figure 6. Student Practicum for DePauw University, 2003. Survey of the Hanna Street Cemetery.

USDA GeoSpatial Data Gateway.

The practicum has several stages, and throughout their experiences, the campus teams report weekly on their progress to their peers and instructor through the CDS. Though we break down the stages into distinct tasks for the sake of project management, we work on the practicum throughout the semester. Issues related to the practicum are interwoven into other segments of the class, and students apply their growing knowledge to their own practicum projects, which they complete by the end of the semester.

Stage 1: Formulate research questions, choose an area, and develop a survey strategy. Each group decides on particular social questions that they might be able to answer through the collection or recording of artifactual evidence. They then determine the limits of their survey region. The group also decides on a sampling strategy.

Stage 2: Develop a collection strategy and design a database. Groups decide on a collection strategy and design data sheets for each field as well as the entire computer database.

Stage 3: Conduct the survey. Students implement the survey they have designed using GPS units to collect locational data, and using their data sheets to collect attributes for the fields in their database (either in the field or using local archives and libraries). They then report back to the class on their field experience: what worked and what did not?

Stage 4: Map the results. After its data is collected and entered, each group creates a preliminary hand-drawn map of its survey area. We consider how to render topography, what physical structures require inclusion (e.g., buildings, trees, and other elements that may not be the object of inquiry), how those structures and the data points would best be symbolized, and how scale affects the impression of data patterns. The point of this exercise is to prompt the students to conceptualize and render their study area and their data in cartographic terms.

Figure 7. Student practicum for DePauw University, 2004. The sacred landscape of Greencastle.
USDA GeoSpatial Data Gateway.

Stage 5: Integrate the survey map and the database in a GIS. After their brief immersion with ArcGIS software during the midterm meeting, students begin to build a GIS for their survey. They note both the problems they encounter as well as new questions that come up during the process.

Stage 6: Visualize, explore, and analyze patterns and trends. Using GIS as both a tool and an environment, the students visualize, manipulate, explore, and test the data, progressing toward resolution of their initial research question(s) and eliciting additional questions.

Stage 7: Report on the process and principal results of the practicum. Because all archaeologists are responsible for disseminating their results, each group prepares an oral and written report for other members of the seminar. The report summarizes their practicum from start to finish. It must explain the research question, methods, and data (what are the significant patterns?), and offer preliminary interpretations of that data (what might the patterns mean in the context of understanding human activity, society, and the landscape?). This reporting is meant to model the principal modes of scholarly exchange, such as publications in journals and presentations at professional meetings.

Ultimately, this practicum accomplishes two goals: the technical goal of introducing undergraduates to GIS software, and the research goal of engaging students with the primary questions posed by archaeologists in landscape studies. Even though their field work is local and not based in the Mediterranean, the opportunity to conduct their own field survey work provides them with a context for working with others' findings in the remote Mediterranean. As a whole, the seminar and its practicum provide the methodological and theoretical background for students to continue with more advanced work in archaeology and GIS, either as part of the CGMA project or on other archaeological field projects.

Students who wish to maintain their involvement with CGMA and MAGIS beyond the seminar may do so through extracurricular activities throughout the year. In the spring, one or two of the seminar students are eligible for work-study at each institution, to continue some of their previously started research. Additionally, CGMA offers one summer research internship on each campus for students who have participated in the seminar. Students work at their home institutions or at DePauw, the home of the project, and they earn a stipend for their efforts.

Lessons learned

Teaching a multi-institutional course that contains a field component and an online learning component, as well as a connection to the ongoing construction of a major research database, has required us to carefully organize how we manage the course. ICC courses are not typical at our colleges, but it is clear that the students benefit from being part of a collaborative learning experience and having the exposure to multiple faculty members. Details and logistics have sorted themselves out over time; for example, now we recognize that having the faculty member with primary charge of the course for that year determine the assessment criteria is most efficient.[9] We have learned how to condense faculty training on GIS software so that they are able to apply it to their specific needs as quickly as possible. Our objective has never been to have students demonstrate mastery of a software package. Rather, each time we have a student spatially analyze his or her survey data and share the analyses with maps, we know our approach has been successful.

Since the fall of 2003, twenty-three students have completed the CGMA course, and they have collected and entered data on nearly 200 survey projects. One CGMA alumna is currently working as an intern in the DePauw University GIS Center, and several others have applied their knowledge from CGMA to field projects in North America and the Mediterranean. We are looking forward to testing and releasing the first version of MAGIS to the scholarly world, and continuing to synergize teaching and research of GIS and archaeology for our undergraduate students.

Acknowledgments

CGMA is funded by a grant from the Andrew W. Mellon Foundation and is supported by DePauw University, Millsaps College, Rhodes College, and The College of Wooster. The project is overseen by a board composed of survey archaeologists and technologists with expertise in databases and GIS. We would like to acknowledge and thank our coprincipal investigators: Michael Galaty (Millsaps College), P. Nick Kardulias (The College of Wooster), Kenny Morrell (Rhodes College), and all of the students at our schools who have helped advance the project. We would also like to thank M. Beth Wilkerson (DePauw University), our CGMA programmer, and Alexander Iliev (DePauw University), who has served as a student technology intern for two years (funded by the ITAP program at DePauw).

165

Notes

1. Each year more undergraduate courses in classical archaeology incorporate class sessions on GIS, but few offer hands-on training. Increasingly graduate programs in classical archaeology are including GIS as part of their curricula.

2. Sunoikisis, the Associated Colleges of the South (ACS) collaborative program in Classics. www.sunoikisis. org/.

3. ACS (Associated Colleges of the South) Archaeology Program. www.colleges.org/~turkey/.

4. For MAGIS, we have chosen open-source, nonproprietary software: MySQL (Structured Query Language) for the database, GRASS (Geographic Resources Analysis Support System) for the GIS and MapServer for the GIS Web interface. The development hardware is an Apple Macintosh Powerbook G4 running OS X; the server hardware is an Apple XServe with a 2-terabyte RAID array, running OS X. See www.magis. depauw.edu.

5. The Greater Mediterranean is both a cultural and geographic term. It includes areas incorporated within or with close connections to the civilizations centered in the Ancient Mediterranean, i.e., from Morocco to England in the West, and from the Persian Gulf to the Caucasus in the East. Any boundaries are to some extent arbitrary, and we may expand their reach in the future if necessary. MAGIS began with the GTOPO 30 data with 1 km resolution, but we are now implementing the much more detailed SRTM (Shuttle Radar Topography Mission) data, with 90 m resolution (SRTM) Data Server. srtm.csi.cgiar. org/).

6. Using the Spatial Search page (figure 1), the user is presented with graphical tools to poll the database for information. Users may select from the entire Mediterranean Basin or zoom in on a specific location. Individual survey boundaries are displayed as well as a list of projects in a given region. Various options are available for overlays and labels (e.g., names of survey projects, major ancient cities, rivers/lakes/streams, modern political districts, ancient regions). The Database Search page (figure 2) allows the user to query the database for a list of surveys that match given criteria, or browse by several fixed categories. The user may enter or select search criteria for any combination of the database fields, including Survey Name, Principal Investigators, Dates of Operation, Chronological Periods, Methodology, Environment, and Bibliography.

7. ACS Course Delivery System (CDS). cds.colleges.org.

8. We are using ESRI ArcGIS software with ArcView functionality and the Spatial Analysis extension. First we conduct a review of projection and coordinate systems (concepts already introduced in class), followed by essential setup features for ArcGIS, such as data frame properties and relative path names. Then we walk the students through the creation of a new ArcMap software document, covering raster and vector data types, layer superposition, importing publicly available data, creating new shapefiles and associated tables, and simple query and display operations.

9. These have included participation (in class, on the CDS discussion board, and "live" at the midterm workshop); group activities, (i.e., the practicum, which is graded in stages as students report on and post evidence of their progress to the CDS); individual student contributions to building MAGIS; and a take-home final exam. In assessing the practicum, as much weight is given to the process as to the final product, and the take-home final asks students to apply methods and theories of landscape archaeology discussed in the course to a particular case study.

References

Alcock, S., and J. F. Cherry, eds. 2004. *Side by Side Survey: Comparative Regional Studies in the Mediterranean World.* Oxford: Oxbow.

About the authors

Pedar W. Foss and **Rebecca K. Schindler** are both associate professors in the Department of Classical Studies at DePauw University in Greencastle, Indiana, where they teach courses in classical archaeology, Latin, and classical civilizations. Together they codirect the Collaboratory for GIS and Mediterranean Archaeology (CGMA). Over the last decade, they have used GIS in the context of their archaeological research: for Rebecca's work on Greek sanctuaries and the ancient landscape, and for Pedar's survey of the Elmali Basin in southwest Turkey.

Teaching the natural sciences with GIS

Introduction to chapters 12–15

Diana Stuart Sinton

Natural phenomena, in one form or another, are at the core of the natural sciences, and data is used to model and represent these phenomena. Whether the discipline is chemistry, geology, or biology, scientists collect information to evaluate a hypothesis, interpret the data, and draw conclusions. They are often interested in how a phenomenon has changed over time and may collect and manage datasets that are numerous and large. They are curious to know how an outcome will vary when parameters are altered, so they build models. They want to study processes and patterns that occur at landscape and global scales, so they use images collected by airplanes and satellites. They want to see patterns that our human eyes can't detect, so they use alternative sensors such as infrared imagery and underwater sonar-based imagery. Finally, they share their findings with others, so that their disciplines progress toward a greater collective understanding.

With the development of desktop GIS, more faculty and students at undergraduate institutions are finding mapping technologies accessible and affordable. At the same time, through the growth and expansion of the Internet as a common means for data dissemination, we have easier access to reams of spatial data, especially from federal and state agencies.[1] A second concurrent development has been the dramatic rise in popularity of handheld global positioning system (GPS) receivers. When President Bill Clinton turned off the government's intentional degradation of GPS signals for civilian purposes (known as "selective availability") on May 1, 2000, he opened the door for millions of students to collect accurate locational data with affordable receivers.[2]

For all of these reasons and more, GIS effectively supports teaching and research in the natural sciences. Whether the objects of study are trees, ponds, birds' nests, rock slopes, or wetlands, science students sample data, measure quantitatively (for example, diameter, pH, density, angle, extent), and may mark locations with GPS to revisit later. GIS mapping imposes a bird's-eye perspective of looking down on the subject of interest, and natural scientists are accustomed to discerning patterns from that perspective (arguably more than humanists and many social scientists, for example). While they may see the patterns, they study the processes that generate them, and GIS enables scientists to visualize, question, and explore pattern and process interactions.

GIS and other mapping software enable the visualization of spatial information and are embedded within the more general field of scientific visualization.[3] Faculty integrate visualizations into their courses because they believe it helps them explain the material more fully and helps the students learn it more comprehensively.[4] In this volume, we hear faculty from across all disciplines say that students in today's classrooms engage most readily with visual material.

GIS' capacity to handle large and multiple datasets facilitates its use for monitoring changes over time in those data. Biology students in Ray Heithaus's class at Kenyon College contribute annually to a long-term database of ecological information from a nearby forest. Chemists Brigitte Ramos and Shelie Miller joined with environmental scientist Karl Korfmacher to map amounts of sedimentation in a pond by identifying the location and quantity of lead present in those sediments.

Geologist Suresh Muthukrishnan considers change over time within a framework of modeling. His students process satellite imagery to quantify land-use changes over time and use that information to model the impact of urban development on water quality. He feels strongly that when his students know the science behind these models, and understand how their predicted results are dependent on zoning regulations, they can become informed and active citizens in a planning process. Models also play a role in Colby College's Problems in Environmental Science course, as students predict the potential for erosion into a pond by mapping and evaluating contributing factors such as slope, soil type, and land-cover data. GIS-based models support learning by allowing students to tweak parameters and have the immediate gratification of visually appraising the results.

Today's undergraduate curriculum in the natural sciences focuses on inquiry-based learning and real-world problem solving. Field work is a common component of courses, and students may work side-by-side with faculty to conduct primary research. In fact, the line between research and experiential coursework frequently blurs, particularly in upper-level classes. Faculty create for their students learning experiences that are as real-world as possible, such as the environmental assessments of aquatic ecosystems conducted at Colby College and the assessments of restored wetlands at Kenyon College. Moreover, undergraduate research has the potential to be just as complex, substantial, and thorough as that conducted by graduate students or other researchers, such as Shelie Miller's work in chemistry.

GIS also has a role in the dissemination of data and research results, enabling students to interact with other scientists in their disciplines. Robert Mauck's biology students at Kenyon have been mapping the distribution of bird nests on a remote Canadian island, interacting with other biologists and contributing to the collection of long-term data for the bird population. Research in Colby's Problems in Environmental Science course addresses issues raised by a local community group or state agency. Students' GIS analyses are considered thorough and professional by the groups with which they work.

These four case studies illustrate why GIS has a place in undergraduate science classrooms. Students gain hands-on experience collecting their own field data, a process that establishes important connections to their local places. Other connections may develop as well, between the students, their college, and local community groups. The research embedded into each of these courses matches the activities of real-world scientists, though in each case GIS is only one of many tools within these natural science disciplines. When spatial data is the focus, GIS supports the whole learning process, from data organization to analysis, predictive modeling, and sharing those modeled results as easy-to-interpret images. GIS helps natural science students question the patterns and the processes.

Notes

1. The United States Geological Survey, the Environmental Protection Agency, the National Oceanic and Atmospheric Administration, the National Park Service, and the Bureau of Land Management, to name a few.

2. Selective Availability (SA) is a feature of the Global Positioning System by which the Department of Defense can deliberately degrade satellite signals to reduce a signal's accuracy. This feature was created for the sake of national security. Without SA, accuracy may be up to ten times better. More expensive receivers and their software packages can accommodate differential "correction" of degraded signals, so the biggest gain has been seen with the lower-end units.

3. See for example, www.nsf.gov/about/history/nsf0050/visualization/future.htm and svs.gsfc.nasa.gov/.

4. See chapter 1: Critical and creative visual thinking (this volume), and examples of these course-based visualizations at serc.carleton.edu/NAGTWorkshops/visualization/index.html.

At Kenyon College we have incorporated GIS into our biology undergraduate curriculum as a tool to promote active learning. We use GIS for all biological inquiries, emphasizing students' creativity and critical thinking. Students apply GIS at spatial scales ranging from individual forest stands to entire river basins, and at temporal scales from lab exercises to multiyear, long-term ecological investigations. In this chapter we will describe projects that we have worked on with students, either in class or as faculty-guided undergraduate research.

Chapter 12

Biology: Spatial investigations of populations and landscapes

M. Siobhan Fennessy, E. Raymond Heithaus, and Robert A. Mauck

GIS helps to bring the out-of-doors into the biology classroom, letting students visualize the complexity of natural systems and make connections beyond the confines of the traditional classroom (Friedrich and Blystone 1998). Using GIS, students critically consider the many interrelated factors that influence the abundance and distribution of organisms, communities, and ecosystems. At its best, GIS lets students tell stories, connect their studies with their world, and ultimately generate questions that can be tested though the strategic design of field sampling. Furthermore, pairing GIS with global positioning systems (GPS) means enriching field biology with the power of mapping and spatial analysis.

In our undergraduate biology curriculum, we use GIS to promote active learning. We do not emphasize GIS software skills, which are outside our curricular goals, but we do expect students to achieve a level of competency and confidence. Nor are the faculty experts in the software, but we also believe in using it ourselves, along with the students, so we ourselves stay current and competent to guide them. We emphasize the benefits of GIS as a tool for analysis, an opportunity to grapple with issues of data quality, and a chance to learn the power of visual presentation. GIS serves as a learning environment for all aspects of biological query from the formulation of questions, to data collection and analysis, to the presentation of final results.

In this chapter we describe how we have used GIS in projects that span a range of spatial scales (forest stand, island-based seabird populations, major river basin) and student effort (from class projects to student-based research projects to large modeling efforts with associated field studies involving teams of student researchers). These projects illustrate how biologists' use of GIS has matured from data organization and mapping to spatial analysis and landscape modeling. Our students download and manipulate available digital map data, collect GPS data in the field, and use other standard biological sampling techniques. The use of sampling techniques figures prominently in our pedagogical approach. Students get hands-on experience linking discrete field locations, GPS coordinates, and ecological attributes. Students create a visual representation of a place (a map) that can be stored, manipulated, and displayed using GIS.

Our case studies describe three projects. The first, directed by Ray Heithaus, uses a GIS model to illustrate how environmental factors affect the distribution and abundance of tree species in a temperate, deciduous forest. The second case study, directed by Robert Mauck, describes how students are using GPS and GIS to create a dynamic "living" map of a colonial seabird population (Leach's storm petrels, *Oceanodroma leucorhoa*) on an island in Canada's Bay of Fundy. Siobhan Fennessey directs the third, in which we developed a GIS-based watershed model for wetland conservation.

Analyzing forest structure: sampling and visualizing data

Ecologists strive to explain the distribution of organisms, a challenge when so many different environmental factors can influence species and each species responds individually. GIS allows us to visualize this complexity and helps students to think critically, beyond "one-factor ecology." In addition, GIS is useful for designing and carrying out sampling in the field.

Our advanced ecology lab course serves as a gateway to the study of natural history. Using GIS, students study the environmental factors that contribute to the spatial variation in forest diversity. Each year, students merge the field data they collect with other existing datasets and model forest diversity. Thus, as more study plots are added to the model each year, we refine our understanding of the forest system.

During the three-hour-per-week lab component of this course, students learn sampling and experimental design, data analysis, and other ecological research techniques. We design the GIS-based activities to introduce students to basic concepts behind GIS: many different types of data from varying sources can be included in a model; this data can be layered to be seen coincidently and simultaneously; we can search for and identify patterns. All of our GIS activities (acquiring basic software skills, becoming familiar with GIS concepts, establishing spatially randomized sampling points, and exploring the visual patterns of forest diversity) occur within the course framework of analyzing and interpreting forest structure.

Students conduct field work at Kenyon College's Brown Family Environmental Center (BFEC), a 380-acre facility with diverse terrestrial habitats.[2] The center's close proximity to campus creates an excellent opportunity to visit repeatedly and accumulate long-term data. Forest habitats vary from early successional to mature, mixed-age stands. In addition, two managed areas were planted with white pine (1,000 trees in two spatial arrays) and a hardwood riparian restoration (3,000 trees of seven species).

Starting with the technology

Students work independently with ArcGIS using the ArcCatalog preview to explore the vector and raster files. Our baseline datasets include soils, road and stream networks, ponds, trails, elevation contours, and aerial photography.[3] Students also have access to a layer created by previous classes that depicts study plots of forest diversity.

Within ArcMap, students assemble and review the data layers and their attribute tables, observing and exploring interrelationships within the data. For example, they compare classification of the twenty-two different soil types to the slopes indicated by the elevation contour layer. Students are asked to predict how forest composition and structure might vary through the BFEC, given these environmental variations. Students also examine the distribution of study sites from previous classes and select new study areas based on their hypotheses.

Using GIS to establish sampling points

Gathering information from certain selected trees is an important step since we cannot sample the entire forest. We use point-quarter sampling to measure tree diversity within forest plots, and students calculate importance values (an index based on the frequency, aggregation, and size of individuals) for each species of tree. The sampled trees must be selected randomly within each plot to support our use of statistics, and the plots themselves must be situated a minimum distance apart from each other to reduce the probability of resampling individual trees. An ArcGIS script facilitates our sampling scheme, and students can print maps of study plots to use in the field.

Using GIS to visualize patterns

After the fieldwork, students combine their sampling results to create a class dataset and analyze their collective data. They create an inventory of sampled tree species that includes the number of individual trees and importance values for each species. Their results are then merged with the layer of sample plots collected in other years by previous classes. Students generate maps that illustrate abundance differences for particular tree species that represent early successional, late successional, or riparian species (figure 1). They also create aggregate indices for species that are especially shade tolerant or that have seeds dispersed by wind. Students then visually explore differences in forest composition as a function of variation in soil type, slope, and the history of disturbance for different sites. As students contribute to this database, observing and recording multiple factors, and examining the relationship between factors, they engage with the material with more enthusiasm and diligence than we observed previously. Being able to be involved with all aspects of the study, and visualizing their research through maps, catalyzes their enthusiasm. Each year, as more study plots are added to this GIS model, we will have the advantage of refining our understanding of environmental factors that contribute to spatial variation in forest diversity.

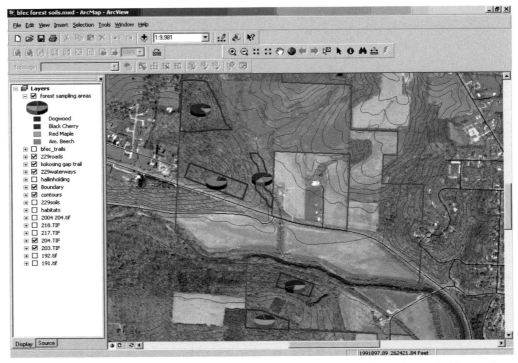

Figure 1. Screen image showing study plots outlined in green and the relative abundances of two colonizing species (red maple and black cherry) and two shade-tolerant species (American beech and flowering dogwood). The boundary of the Brown Family Environmental Center at Kenyon College is red; contour lines give 20-foot intervals.
Aerial photo by Knox County Ohio Auditor's Office; data from USGS.

Mapping islands and population ecology with GPS and GIS

Marked populations of wild birds are rare and long-term monitoring of such populations is even more so. The breeding population of Leach's storm petrels at the Bowdoin Scientific Station at Kent Island, New Brunswick, Canada, has been monitored continuously since 1954. Leach's storm-petrels (figure 2) are small (45 grams), pelagic seabirds (order Procellariiformes) that breed on islands in the northern hemisphere and make their nests in holes dug into the forest floor (Huntington et al. 1996). Once dug, these burrows persist, with individual birds often returning to the burrow they occupied the year before. The Kent Island population is a remarkable resource for addressing our questions related to age-specific phenomena because we have complete breeding histories for individual birds, some of which have bred for thirty years. There are few, if any, comparable long-term projects in the world. It is a system ripe for collaboration; however, until recently, collaboration was curtailed because there was no readily accessible and sharable source of spatial data.

The Bowdoin Scientific Station is located on Kent Island, New Brunswick, Canada (44° 35′ N, 66° 46′ W), 9 kilometers south of Grand Manan Island at the mouth of the Bay of Fundy. The storm petrels nest primarily on the northern half of the island in a spruce, fir, and mixed hardwood forest.

Figure 2. Leach's adult storm petrel.

Dr. C. E. Huntington of Bowdoin College monitors about three hundred to four hundred pairs annually in three noncontiguous study areas on the island. For several years, Robert Mauck has involved his Kenyon students in the collaborative research he conducts with Huntington to study the breeding biology of this seabird. Through the years, Huntington has identified and monitored more than 1,200 petrel burrows. Each year, new burrows are found and a few old burrows go vacant. Historically, Huntington numbered each newly found burrow chronologically, with the new burrow simply receiving the next incremental burrow number. Yet no explicit locational information was embedded into each burrow's naming convention. Instead, burrow locations were originally documented in a notebook with descriptions such as "2m SE of [burrow] 325 and 6m N of [burrow] 1988." Such information is of little use unless the corresponding reference burrows can be found. Additionally, if the reference burrows are no longer active, finding the burrow of interest would be impossible. Thus, our first goal was to locate existing burrows in absolute space and store that information in an accessible format.

Transforming spatial data: differential GPS and the quest for precision

To begin this project, Jenny Glazer (Kenyon 2003) used GPS and GIS tools to create a map of the petrel colony on Kent Island. Jenny used the Trimble Pathfinder ProXRS, twelve-channel, multiple-satellite receiver for real-time differential global positioning system (DGPS) data acquisition, more accurate than less-expensive handheld devices. We needed the submeter precision of DGPS to create reliable maps that would lead the student researchers to nesting burrows, to record the necessary demographic data.[4] Also, such precision would enable any subsequent analysis of interannual movement patterns within the colony. During the normal censusing of the petrel colony, Jenny used the Pathfinder to determine GPS locations as Huntington visited each nesting burrow. She also mapped every path and important landmark in the colony. Using ArcGIS, we printed updated maps and used them to double-check the accuracy of the GPS information, then adjusted the digital map accordingly.

Our main challenge was working in the forest. Petrels nest on the forest floor yet canopy cover causes multipathing errors when GPS signals are reflected off trees or are completely blocked by the forest canopy. Such errors can degrade precision by one to five meters, and it was critical that we map the burrows with as much precision as possible (Johnson and Barton 2004). To overcome these effects, we recorded points only when a minimum of six satellites were available and their geometry provided an adequate amount of coverage throughout the sky (i.e., the level of precision from the signals was not diluted or reduced).[5] We also used a 10-foot extension pole to raise the GPS antennae high into, or above, the canopy (figure 3), and collected multiple readings (at least

Figure 3. Kenyon College student using the Trimble Pathfinder with antenna extension to map petrel burrows under the forest canopy.

Photo by Robert A. Mauck.

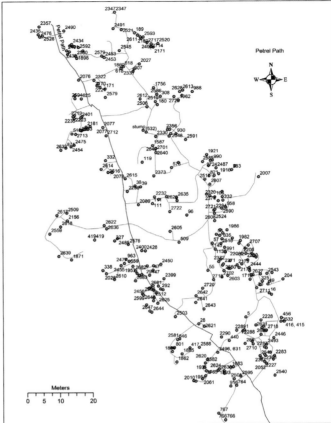

Figure 4. Large-scale map of the petrel path study area on Kent Island.

Data collected by Jenny Glazer.

twenty) for the same point to increase our precision. We found, in most cases, such redundancy sufficiently improved the precision to justify the extra effort.

Through conscientious GPS work in the summer of 2002, we mapped more than 300 burrows, important landmarks, and walking trails, with a mean horizontal precision of 0.55 ± 0.20SD meters. Jenny began with spatial information keyed to one researcher's memory of nest locations, and converted it into easily accessible and maintainable maps (figure 4).

Communicating results

As the mapping effort wound down, the new map was put to the test. A team of ornithologists arrived to sample blood from specific, known-age petrels. The team gloomily anticipated days of searching for the target burrows, deciphering the convoluted location descriptions. However, using the new GIS, Jenny quickly printed a map complete with up-to-date paths, landmarks, burrow numbers, and annotations. The ornithologists simply walked from burrow to burrow collecting samples. Through GPS and GIS technology and the effort of one student, the petrel colony at Kent Island had been opened to the research world. In the process, Jenny learned to use this important technology for ecologists and gained valuable experience dealing with the real-world challenges of fieldwork. With clearly definable and attainable goals, the project encouraged a genuine sense of ownership along with a well-deserved sense of pride.

Now that the core of the petrel nesting areas has been mapped, students can begin to ask questions. Do petrels nest randomly across the landscape, or are they clumped? Does reproductive success have a spatial component? Does proximity to neighbors matter? Are there "hot spots" within the study area where reproductive success is particularly high? Linking the GIS data to the historical breeding records will help answer these questions. Every question becomes a hypothesis with a potential project for students wanting to explore the power of geographical information systems. In subsequent summers other undergraduates will expand the project to other parts of the island, update the existing maps, locate breeding birds, and continue the demographic study of petrel breeding biology. In addition, the new spatial data can be combined with more than fifty years of demographic data, enabling us to investigate the intricacies of temporal and spatial breeding patterns within the colony.

Spatial analysis, modeling, and fieldwork for wetland restoration

Wetlands serve multiple functions, including flood storage, water purification, nutrient retention, and the provision of wildlife habitat, all of which support the environmental quality of the landscape (Costanza et al. 1997; Nagule et al. 2001). Despite their valuable ecosystem services, wetlands are among the most degraded ecosystems on earth with losses estimated at more than 90 percent in many parts of the world, including Ohio (Mitsch and Gosselink 2000). These losses have prompted widespread wetland restoration programs by both government agencies and nongovernmental organizations.

Efforts to restore wetlands are usually site-specific (National Research Council 2001), and may ignore the broader landscape in which the wetland will be placed (Kershner 1997). How effective a wetland can be at influencing watershed-scale processes, such as water quality improvement and flood peak attenuation, are determined in part by the spatial configuration of those wetlands. Thus planning wetland restoration projects at the watershed scale can yield dramatic benefits. For instance, DeLaney (1995) noted that watersheds comprised of 5 to 10 percent wetland area could provide a 50 percent reduction in peak flood period compared to watersheds without wetlands and significantly improve water quality in those watersheds (Johnston et al. 1990; Day et al. 2003).

Developing a GIS model

Wetlands research has been one of our priority areas for several years, and several faculty had developed a GIS-based model to predict the suitability for wetland restoration, in this case in the

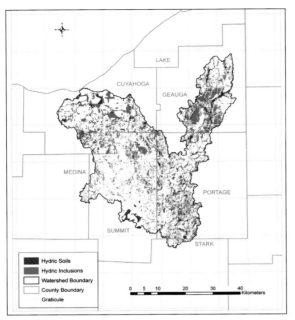

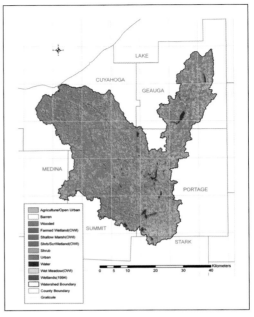

Figure 5. Map of hydric soils and soils with hydric inclusions in the Cuyahoga River watershed (CRW). For the soil series in the CRW, the percentages of hydric inclusions ranged from 2 to 15 percent, with most values at 5 percent.

Reprinted from *Ecological Engineering* 24 (2005), with permission from Elsevier.

Figure 6. Map of land-use land cover in the CRW. Wetland areas are as defined in the Ohio Wetlands Inventory.

Reprinted from *Ecological Engineering* 24 (2005), with permission from Elsevier.

Cuyahoga River watershed (CRW) (2,107 km²) in northeastern Ohio (White and Fennessy 2005). In this model we combined variables important to the success of wetland restoration projects (soil type, land use, hydrology) to prioritize restoration activities. Our goal was to develop a model that used the analytical capabilities of GIS to identify wetland restoration sites in a spatially explicit way.

We used a two-phase approach to create the model. The first step was to develop criteria, or environmental indicators, to identify all sites suitable for wetland restoration (combining raster data layers with a 30-by-30-meter pixel size). We identified locations where restoration projects would have a high likelihood of success and would be sustainable over the long term. Our criteria included hydric soils (figure 5), type of land use (including existing wetlands as mapped by the Ohio Wetlands Inventory, or OWI) (Yi et al. 1994) (figure 6), topography, stream order, and a saturation index based on slope and flow accumulation in each grid cell in the model (White and Fennessy 2005).

In the second phase of our model development, we prioritized the list of suitable sites according to their potential to contribute to downstream water quality. Using a digital elevation model (DEM) and an overland drainage network for the watershed, we determined overland flow directions by searching for the direction of steepest descent from each grid cell in the DEM (figure 7) (Jensen and

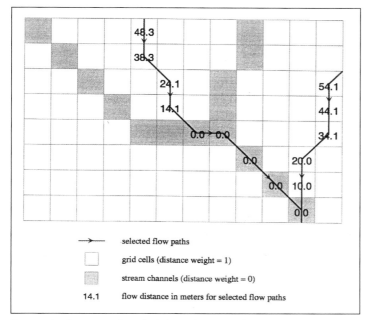

Figure 7. An illustration of how hydrologic distance to the nearest stream channel is calculated. Flow paths are traced downslope from each cell and the distance from the center of one cell to the center of the next downstream cell is calculated.

Reprinted from *Ecological Engineering* 24 (2005), with permission from Elsevier.

Domingue 1988). Once we know the direction of overland flow throughout a DEM, the GIS can then easily determine the amount of flow accumulation in any cell. We then assessed data in other map layers, seeking locations with hydric soils and nonurban land use. We combined these factors in an additive raster model showing the spatial distribution of sites ranging from little or no to high restoration potential (figure 8).

Working in the field and landscape planning

Our overall pedagogical goal was to involve students in watershed-level planning and assessment of wetland ecosystem restoration. Because the model includes both existing wetlands and sites identified as suitable for wetland restoration, faculty adopted it as the basis for a large-scale undergraduate research project to develop "green infrastructure" scenarios for the watershed (Weber and Wolf 2000). This has entailed two steps: validating the results of the restoration model on the ground by systematically checking model predictions (soil type, vegetation cover, slope, and so forth), and performing an assessment of the ecological health or condition of existing wetlands using a rapid assessment method.

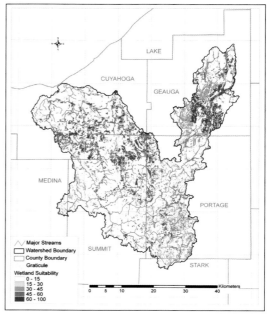

Figure 8. The final model of site suitability of wetland restoration potential.
Reprinted from *Ecological Engineering* 24 (2005), with permission from Elsevier.

Student work began by assessing the existing wetlands using the Ohio Rapid Assessment Method, or ORAM (Mack et al. 2000; Mack 2001). This method is designed to take less than two hours per site during a single visit, and has been extensively calibrated and validated using detailed data on macrophyte, macroinvertabrate, and amphibian communities (Fennessy et al. 1998; Mack et al. 2000). We trained student field crews to locate mapped wetlands using GPS units, determine whether a wetland was actually present at that location, and perform the rapid assessment. Students gained valuable skills in translating mapped locations to real sites on the ground and documenting the attributes of that site.

The rapid assessment method uses structural variables (e.g., the number of plant community classes, hydrologic modifications such as dikes or ditches, the presence of amphibian breeding pools) to identify high-quality or significant wetlands (for example, because they provide important amphibian habitat, or have a high degree of ecological integrity). Those wetlands that we deem most critical for preservation will be integrated with the potential restoration sites to serve as a blueprint for a future "greenway" network of critical wetlands. Such a network is possible through land acquisition, the provision of buffers, and by linking adjacent wetlands into corridors or complexes.

The logistical issues in a project such as this are substantial. For instance, the Cuyahoga River watershed has thousands of wetlands and not all can be visited. Our goal has been to work on two hundred randomly selected sites, perform the rapid assessment at each, then extrapolate the results of this sample to the entire population of wetlands with a known level of confidence. We can do this using a geospatially balanced, stratified random sample by dropping sample "points" onto existing wetlands as identified by the Ohio Wetland Inventory (shown in land-use cover, figure 6). Sample points are given as latitude–longitude coordinates using the Generalized Random Tesselation Stratified (GTRS) survey design (developed by the USEPA Environmental Monitoring and Assessment Program) (Stevens 1997; Stevens and Olsen 1999).

Figure 9. Aerial photo showing an example of a wetland sampling point dropped randomly in the Cuyahoga River watershed. Sampling points were ordered numerically and were drawn from wetlands mapped by the Ohio Wetlands Inventory (white areas). Both 60-m and 100-m-diameter buffer areas are shown around the point.

Image courtesy of M. T. Sullivan.

Together, students and faculty performed a preliminary examination of sample points remotely using U.S. Geological Survey (USGS) digital aerial photos, including digital orthophoto quarter quad (DOQQ) images and high-resolution airphotos. We marked each sample location on the aerial photo and made note of the underlying land use (figure 9). Student field crews then used the ORAM method to sample any wetland located at the point, or within a 60-meter radius of the point. Thus students develop a landscape perspective to generate questions about the effects of changing land use on wetlands. Does land use surrounding a site affect its ecological condition as measured by ORAM? What size buffers appear adequate to protect wetlands? Does wetland density in an area correlate with species diversity? Students are tackling these questions as part of their independent research programs.

Ultimately, our objective is to protect wetlands and stream corridors that are at risk due to human land-use changes. Students will gather information via site visits, close observation of aerial photos, and GIS modeling. The project is designed to integrate GPS and GIS technology with student field research to identify locations where restoration efforts and conservation practices can be applied. In so doing, we hope to improve the environmental health of the watershed and give students the opportunity to participate in important research on conservation planning.

Concluding thoughts

Using GIS in biological studies gives students a unique opportunity to think spatially, present and communicate visually, enhance observation, and develop an appreciation for the complexity of environmental processes, particularly when planning watershed restoration. It also affords our students the chance to grapple with issues of data quality—error, accuracy, and precision—as they did when striving to overcome the challenge of tree cover with multiple GIS readings. In all projects students became familiar with various types of technology, data, and data-handling techniques. Our use of GIS promotes active learning at all levels of the curriculum. It offers a valuable complement to more traditional field techniques in field biology for understanding ecological processes in the context of real-world, spatial complexity.

Notes

1. Many scripts can be found at arcscripts.esri.com.
2. Brown Family Environtmental Center. www2.kenyon.edu/Bfec/.
3. The GIS model for the BFEC includes recently revised layers for soils (provided by the Soil and Water-Conservation District) and aerial orthophotographs (2 feet per pixel, 2000 and 2004 flight data, provided by the Knox County, Ohio Regional Planning Commission). Data layers from the USGS indicate roads, streams, ponds, and elevation contours. The road and stream layers in the BFEC were edited to achieve a finer resolution, consistent with the orthophotos.
4. Differential Global Positioning System (DGPS) uses fixed reference beacons to correct calculations based on satellite signals enabling a high degree of accuracy and precision. These continuously operating reference stations (CORS) are available through government and private sources. CORS coverage on Kent Island benefits from the maritime fishing industry's heavy use of the technology in and around the Grand Manan archipelago.
5. Although one can predict optimal satellite geometry with readily available online software (e.g., Trimble Planning Software), we let the device's Position Dilution of Precision (PDOP) reading tell us when we could map with adequate precision. We determined that readings under 6.0 were acceptable and readings under 4.0 were optimal.

References

Costanza R., R. d'Arge, R. de Groot, S. Farber, M. Grasso, B. Hannon, K. Limburg, S. Naeem, R. V. O'Neill, J. Paruelo, R. G. Raskin, P. Sutton, M. van den Belt. 1997. The value of the world's ecosystem services and natural capital. *Nature* 387:253.

Day, J. W. Jr., A. Yanez, W. J. Mitsch, A. Lara-Dominguez, J. Day, J. Ko, R. Lane, J. Lindsey, and D. Lomeli. 2003. Using ecotechnology to address water quality and wetland habitat loss in the Mississippi basin: a hierarchical approach. *Biotechnology Advances* 22:135–59.

DeLaney, T. A. 1995. Benefits to downstream flood attenuation and water quality as a result of constructed wetlands in agricultural landscapes. *Journal of Soil and Water Conservation* 50:620–26.

Fennessy, M. S., B. Elifritz, and R. Lopez. 1998. Testing the Floristic Quality Assessment Index as an Indicator of Riparian Wetland Disturbance. Ohio Environmental Protection Agency Technical Bulletin. Division of Surface Water, Wetlands Ecology Unit. Columbus, OH. www.epa.state.oh.us/dsw/401/.

Friedrich, R. L. and R. V. Blystone. 1998. Internet teaching and resources for remote sensing and GIS. *Bioscience* 48:187–92.

Huntington, C. E., R. G. Butler, and R. A. Mauck. 1996. Leach's Storm Petrel (*Oceanodroma leucorhoa*). In *The Birds of North America* 233, eds. A. Poole and F. Gill. The Academy of Natural Sciences, Philadelphia, PA, and The American Ornithologists' Union, Washington, DC.

Jensen, S. K. and J. O. Domingue. 1988. Extracting topographic structure from digital elevation data for geographic information systems analysis. *Photogrammetric Engineering and Remote Sensing* 54:1593–1600.

Johnson, C. E. and C. Barton. 2004. Where in the world are my field plots? Using GPS effectively in environmental field studies. *Frontiers in Ecology and the Environment* 2:475–82.

Johnston, C. A., N. E. Detenbeck, and G. J. Niemi. 1990. The cumulative effect of wetlands on stream water quality and quantity: a landscape approach. *Biogeochemistry* 10:105–141.

Kershner, J. L. 1997. Setting riparian/aquatic restoration objectives within a watershed context. *Restoration Ecology* 5:15–24.

Mack, J. J. 2001. Manual for Using the Ohio Rapid Assessment Method, Version 5. Ohio Environmental Protection Agency Technical Bulletin 2001-1-1. Division of Surface Water, Wetlands Ecology Unit. Columbus, OH. www.epa.state.oh.us/dsw/401/.

Mack, J. J., M. Micacchion, L. Augusta, and G. Sablak. 2000. Vegetative Index of Biotic Integrity for Wetlands and Calibration of the Ohio Rapid Assessment Method for Wetlands, version 5. Ohio Environmental Protection Agency Technical Bulletin and Report to U.S. EPA. Division of Surface Water, Wetlands Ecology Unit. Columbus, OH. www.epa.state.oh.us/dsw/401/.

Mitsch, W. J. and J. G. Gosselink. 2000. *Wetlands.* New York: Wiley and Sons.

Nagule, E. D., R. R. Johnson, M. E. Estey, K. F. Higgins. 2001. A landscape approach to conserving wetland bird habitat in the prairie pothole regions of Eastern South Dakota. *Wetlands* 21:1–17.

National Research Council. 2001. *Compensating for Wetland Losses Under the Clean Water Act.* Washington, DC: National Academy Press.

Stevens, D. L. Jr. 1997. Variable density grid-based sampling designs for continuous spatial populations. *Environmetrics* 8:167–95.

Stevens, D. L. Jr. and A. R Olsen. 1999. Spatially restricted surveys over time for aquatic resources. *Journal of Agricultural, Biological, and Environmental Statistics* 4:415–28.

Weber, T. and J. Wolf. 2000. Maryland's green infrastructure: Using landscape assessment tools to identify a regional conservation strategy. Environmental Monitoring and Assessment 63:265–277.

White, D. A. and M. S. Fennessy. 2005. Modeling the suitability of wetland restoration at the watershed scale. *Ecological Engineering* 24:359–77.

Yi, G-C., D. Risley, M. Koneff, and C. Davis. 1994. Development of Ohio's GIS-based wetlands inventory. *Journal of Soil and Water Conservation* 49:24–28.

About the authors

E. Raymond Heithaus is Jordan Professor of Environmental Science and Biology at Kenyon College, where he teaches courses in environmental studies, ecology, and evolution. He developed GIS experiences for laboratories in ecology and introductory biology. He uses GIS as a research and management tool at the Brown Family Environmental Center at Kenyon College, for which he is executive director. He also provides GIS services to a local, nonprofit land trust.

Robert A. Mauck is the Harvey F. Lodish Faculty Development Professor in the Natural Sciences and assistant professor of biology at Kenyon College, where he teaches courses in animal behavior and environmental studies. His work with Leach's storm petrels investigates life-history trade-offs in long-lived animals.

M. Siobhan Fennessy is an associate professor of biology and environmental studies at Kenyon College. She teaches courses in aquatic ecology, ecology, and sustainable agriculture and uses GIS in her research on the effects of land-use change, particularly urbanization, on the ecological health of wetland ecosystems.

Maine is renowned for its abundant forests, lakes, and coastline. Ninety percent of the state is covered in forest, the state's 8,000 lakes and ponds and 32,000 miles of rivers and streams are more than all other New England states combined, and the rugged shoreline stretches an amazing 3,500 miles. Maine's size and environmental diversity represent an opportunity for Colby College students to study environmental science and natural resource management in the field.

Our environmental science program maintains active and collaborative partnerships among faculty, library, and technology staff, and we use this to our advantage to promote inquiry-based spatial learning. We work with students and community partners to study and better understand the state's unique natural resource challenges. GIS has become a central tool for many of these interdisciplinary teaching and research projects.

In this chapter, we illustrate how we are using GIS to facilitate and enhance interdisciplinary problem-based research and teaching in Problems in Environmental Science (PES), a senior capstone course for environmental science majors.

Chapter 13

Environmental studies: Interdisciplinary research on Maine lakes

Philip J. Nyhus, F. Russell Cole, David H. Firmage, Daniel Tierney, Susan W. Cole, Raymond B. Phillips, and Edward H. Yeterian

Since the late 1980s, students in Colby College's Problems in Environmental Science (PES) course have been conducting studies of lakes and watersheds that surround their campus. Colby is located in central Maine, in the center of the state's lake district (figure 1).

The lakes are easily accessible to residents and tourists alike and are estimated to generate over $1.8 billion for the Maine economy each year. Their popularity, however, is contributing to their declining water quality, as housing and road construction have accelerated. A growing number of lakes now experience regular algal blooms resulting from excessive nutrient runoff from surrounding watersheds. This lake degradation has potentially serious implications for water quality as well as for the fishing and recreation sectors of Maine's economy.

Professors Russ Cole and David Firmage designed and teach the PES course to provide a common research experience for environmental science majors and to build on previous coursework in

Figure 1. Students in the Problems in Environmental Science course in the Environmental Studies Program at Colby College studied this map showing the location of lakes in central Maine.
Data courtesy of the Maine Office of GIS, and USGS.

ecology and environmental science.[1] Each year, the class assumes the role of a consulting firm and the students engage in real-world research. They problem-solve collaboratively and have ample opportunities to express themselves orally and in writing. They employ research methods commonly used in the field of environmental science and develop an understanding of relevant state and local regulations and their applications.

This course also exemplifies service learning in action, engaging the college and the environmental studies program with the local community (Firmage and Cole 1999). Which lakes are studied each year depends on incoming requests from local lake associations and the state Department of Environmental Protection (DEP). The course culminates in a public presentation and a major research report of class findings. In previous years, the students' final products have been used by the lake associations to guide management decisions and by the DEP to develop lake management strategies.

One iteration of the Problems in Environmental Science course

In autumn 2004, PES students investigated the water quality and related factors in the watershed of Togus Pond near Augusta, Maine. Situated close to the state capital, residential development along the lake's shoreline has been rapid in recent years and the lake experiences annual algal blooms. Understanding what causes algal blooms was only one component of the PES course, though. Students also investigated shoreline development and surrounding land-use and land-cover change, and modeled erosion potential and site suitability for septic systems. In each case, GIS was used to organize and analyze the environmental data, and we will briefly highlight details of these analyses.

Algal blooms respond to a lake's flushing rate, or its ability to self-clean. To understand flushing rates, the students conducted a descriptive spatial inventory of the lake. Using boat surveys, the class measured the lake's depth at more than 100,000 points and generated a bathymetric map.[2] With bathymetric data in hand, the students could derive vital statistics that described the lake, including its average depth (5.4 meters), deepest point (15.7 meters), and relative distribution of shallow- and deep-water areas.

The students then pinpointed locations in the lake where the depth could jeopardize environmental health or make that location susceptible to an ecological threat. For example, one quarter of the pond is more than 8 meters deep, making it vulnerable to anoxia (a condition when little or no oxygen is available) (figure 2). Anoxic conditions are directly detrimental to many fish species and contribute to the growth of algae (by prompting release of phosphorus from sediments and increased phosphorus loading into the water column, which may stimulate algal growth near the water surface). During summer stratification, the students found anoxic areas in the deepest sections of the lake. The bathymetric analysis also showed that half the lake is susceptible to invasion by exotic macrophytic plants that prefer to colonize water shallower than 5 meters. This type of information is particularly important for scientists and lake association members trying to understand thresholds for lake water quality and working to develop water quality management strategies.

In addition, students conducted a survey of shoreline residences by boat and surveyed the buffer of vegetation that separated each house from the lake. They used GPS to locate the approximate

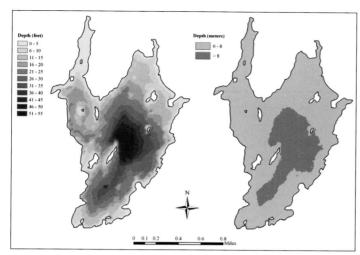

Figure 2. Bathymetry map for Togus Pond with depth measured in feet (left). Depth points were acquired using a Lowrance LCX-15MT, which consists of a boat-mounted sonar device and a GPS unit. The area below a depth of 8 meters (26 feet) where anoxic conditions are common is shown in green (right).
Data courtesy of authors the Maine Office of GIS.

center of each residential lot; and for each house, estimated lot length and buffer slope, vegetative cover, and depth. They made note of dense residential development areas, poor-quality roads adjacent to the shoreline that might contribute to soil erosion, and the quality of shoreline vegetative buffers surrounding the lake (figure 3).[3]

Because changes in land use and land cover have tremendous impacts on water quality, students digitized aerial photographs from 1954 and 2002 to create land-use data layers.[4] The students identified unique land covers for both years, including agriculture, disturbed forest, forest, commercial and municipal, residential, and wetlands, and made these into ArcGIS feature classes (figures 4 and 5).

They used the GIS to calculate the difference between the 1954 and 2002 images and quantified the changes on a map. Like much of the rest of Maine, forest cover has remained relatively stable in the area, accounting for just over 60 percent of the

Figure 3. Buffer evaluation map for Togus Pond. Points correspond to developed house lots evaluated by students using a boat survey of all lakeshore lots.
Data courtesy of the Maine Office of GIS.

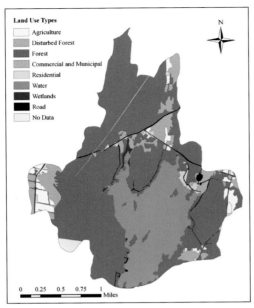

Figure 4. Land-use patterns for the Togus Pond watershed in 1954. Land-use types were determined using aerial photographs. Students digitized roads from the photographs because no road maps were available from that time.

Data courtesy of Colby College 2004 Problems in Environmental Sciences class

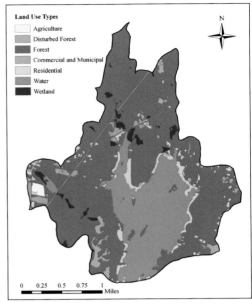

Figure 5. Land-use patterns for the Togus Pond watershed in 2002. Land-use types were determined using aerial photographs. In 2002, students were unable to differentiate mature from transition forest, so only one forest class was used. No cropland was present.

Data courtesy of Colby College 2004 Problems in Environmental Sciences class

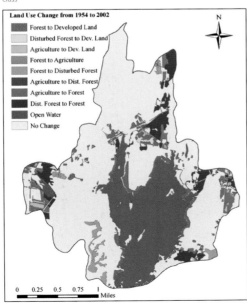

Figure 6. Land-use change in the Togus Pond watershed, 1954 to 2002.

Data courtesy of Colby College 2004 Problems in Environmental Sciences class

total land area. Agricultural land decreased, while residential and commercial land use increased, particularly near the water (figure 6).

Students used the derived data to develop a model that predicted total phosphorus levels in the lake. The model is based on average rainfall, runoff data for different land-use types (adjusted for total area covered by each land-use class), the area and volume of the lake, and the amount of water flowing into the lake over a year's time. The class also projected future phosphorus inputs based on probable future changes in land use, and we plan to make these predictive models spatially explicit in the future.

The students developed a second model that combined four factors (slope, soil type, proximity to the lake, and land-use type) to calculate an area's erosion potential and suitability for septic systems.[5] For erosion potential, the students weighted their overlay model to favor slope as the most significant input (40 percent), followed by soil type (25 percent), land cover (25 percent), and distance from the lake (10 percent). Each factor was classified on a scale of 1 (low) to 9 (high). The final model showed areas with relatively more or less erosion potential (figure 7). They developed a

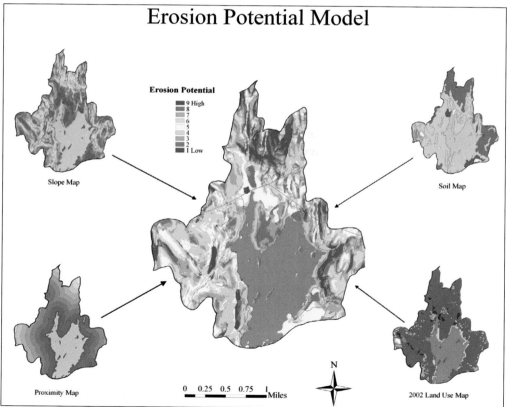

Figure 7. Erosion potential map for Togus Pond watershed derived from slope, soil, land use, and proximity to lake. Purple represents higher erosion potential and green represents lower erosion potential.
Data courtesy of Colby College 2004 Problems in Environmental Sciences class

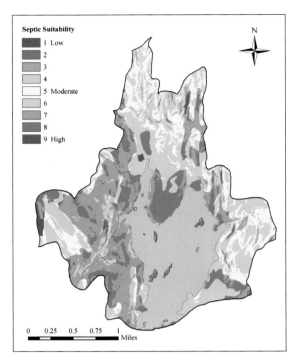

Septic Suitability
- 1 Low
- 2
- 3
- 4
- 5 Moderate
- 6
- 7
- 8
- 9 High

N

0   0.25   0.5   0.75   1
Miles

Figure 8. Results of a model showing septic suitability of the Togus Pond watershed based on soil septic limitations and slope. Areas with low septic suitability are shown in red.
Data courtesy of Colby College 2004 Problems in Environmental Sciences class

similar model, with modifications to the input values, to show which areas were more or less suitable for household septic systems (figure 8).

Impact of GIS on teaching and learning

Collectively, these GIS analyses have had a substantial impact on what and how students learn in the course, and on the impact this course has had well beyond the classroom (Nyhus et al. 2002). Student evaluations have consistently been very positive, and many students have used these experiences, and their GIS skills, to pursue employment and graduate study in related fields. Since the first studies by this course in 1988, students' models have become much more sophisticated, a development that has been concurrent with other progress in our field (such as our ability to measure phosphorus more accurately). The software's usability has increased sufficiently to enable students, some of whom have never been exposed to GIS before, to generate relatively sophisticated GIS land-use models in just one semester (figure 9).

Digital data has become more available and more refined. Increasingly, data is available through the State of Maine and other public sources. Students can download a wide range of thematic layers, such as soil types, rather than painstakingly digitizing the data as they had to do in earlier years. This has enabled students to spend more time on in-depth data analysis and modeling rather than repetitive data preparation. The students, building on methodologies and results of past student projects (available through old reports archived in the laboratory and the library), have a much better perception of what they can accomplish and how important it will be to the Department of Environmental Protection (DEP), the lake associations, and local landowners. This year the DEP incorporated some of the class data (with appropriate credit and acknowledgments) into a Total Maximum Daily Load (TMDL) study they were undertaking.

GIS has improved our ability to communicate effectively and professionally. As the reputation of our course and students' work has became known to state agencies, lake associations, and local land holders, students' expertise has been sought after by citizens and government officials. By 2005, sixteen different lakes had been studied, several of them by other classes or independent projects, and much of

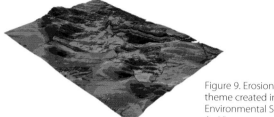

Figure 9. Erosion risk theme draped over topographic theme created in 2000 by students in Problems in Environmental Science class. This was the first year ArcView was used by students in this course.

this research has been enhanced by GIS. The professional quality of the analyses and resulting images created compelling visual displays that engaged the audience and lent credibility to the studies during the class's public presentations. The bathymetry, shoreline, land cover, and suitability maps provided unique insight into the biophysical dynamics of the watershed basin. They highlight high-risk areas that would benefit from short-term and long-term remediation. These conclusions informed not only the students, but also local watershed, regional, and statewide management, planning, education, and policy. Furthermore, students often present their research at local, regional, and national meetings.

Expanding the use of GIS within environmental studies

Our PES course is one of several in the environmental studies program that use GIS and other geospatial technologies (figure 10). Colby College recently hired a new environmental studies faculty member, Philip Nyhus, with expertise in both environmental policy and GIS, adding a strategic resource that benefits several disciplines. A new environmental policy senior capstone course now mirrors the environmental science capstone; both require students to complete a GIS project. For the first time, we are now offering a midlevel course that teaches GIS and remote sensing skills as its primary focus, which is distinctive from other courses that teach with GIS and only apply the tool as needed. This course, designed primarily

Figure 10. Students in Problems in Environmental Science class carrying out water quality studies on a Maine lake.
Photo by David Firmage.

for environmental studies majors but available to students across the campus, allows students to work on projects with community partners and generates a regular series of student Web projects under the broad title of An Atlas of Maine.

Lessons learned implementing campuswide GIS

From its early roots in the environmental studies program and the department of geology, the body of GIS users and the use of GIS in the Colby curriculum continues to expand. Some of the key lessons we have learned from our experience building GIS capacity in our environmental studies program and across campus include the following:

- GIS can promote interdisciplinary teaching, research, and collaboration. Many of the world's most pressing problems are interdisciplinary and are influenced by complex factors operating at multiple scales and levels. In our experience, GIS has been a highly effective teaching and research tool for students to use in investigating local, regional, and global environmental problems. In environmental studies, we have found GIS to be an ideal tool to support studies at a watershed level, because it can incorporate and be used to analyze a wide range of data at varying scales. As the amount and quality of spatial data increases, and as the scope and user friendliness of GIS increases, we expect our ability to use GIS to help students explore these issues will only improve. Moreover, the universal visual language of maps provides a unique opportunity for interdisciplinary curricular initiatives and engagement of students and faculty across departments and academic divisions. The utility of incorporating interdisciplinary GIS analyses into environmental studies inquiry-based courses and some civic engagement courses has been pronounced, and we are replicating the environmental studies model in other curricular initiatives.

- GIS can promote active learning styles. Hands-on activities, inquiry-based learning, and action-oriented investigation represent three successful and highly touted educational approaches, and have each been encouraged by the National Science Foundation, the National Research Council, and the American Association for the Advancement of Science. Emerging geospatial technologies encourage and support these approaches across the curriculum. As one of these technologies, GIS facilitates student engagement in the collection, analysis, and presentation of spatial data. Involving students in real-world problem-solving exercises generates considerable enthusiasm among students, faculty, and staff, as well as outside collaborators and other constituents. Students are motivated to learn and apply GIS—and correspondingly to work hard on the problems they are trying to solve—when the connections are real. Other students in turn are motivated to learn this powerful tool when they see the results obtained by their peers.

- GIS can be used effectively in numerous courses. Interest in GIS is emerging across the campus at Colby. The enhanced functionality of modern GIS software enables students to complete sophisticated projects within a single semester. Undeniably, the learning curve is steep for students with no background in GIS or who are not accustomed to thinking about a problem spatially. But we have found that the abundance of ready-to-use data, increasingly user-friendly software, and publicly available training modules support GIS as a tool that can be used actively in virtually any course across the curriculum. Benefits also exist from a more passive form of use—through increased exposure and access to maps or from the derived products of GIS analyses completed by others. In either case, students benefit by having additional opportunities to think about things from a spatial perspective.

- Faculty and students in the departments of biology, geology, chemistry, government, and economics are now using GIS, and several new courses and research projects that will use GIS are under development. For example, chemistry students are conducting detailed assessments of lakes from which local towns obtain their drinking water. Economic students are working with a faculty member to help the local city government integrate GIS-based census data with local tax and real estate information. Moreover, the GIS activities are not limited to Maine. Students in government and environmental studies are helping faculty members use GIS to better understand spatial factors relating to social justice, civil society, and democracy in Buenos Aries, Argentina. Also, environmental studies students have worked with a faculty member to develop maps of suitable tiger habitat in Indonesia and China. Many of these problems would have been difficult to investigate due to practical and time constraints if GIS resources were not readily available across campus.[6]

- Libraries can and should play important roles in expanding GIS across campuses. Colby's library played a central role in GIS growth on campus. For the PES course, the science librarian prepared a guide presented every semester that identifies sources of spatial data, including datasets, data clearinghouses, and mapping projects. The library has become a central repository for archiving GIS projects, developing and posting subject guides, and maintaining the Colby GIS Web page. The science librarian and the government documents librarian have been instrumental in coordinating a campuswide GIS users' group.

- GIS is just one of many new technologies being introduced to faculty and students. Libraries are uniquely positioned to support these new technologies because they serve all departments and programs on campus and they can leverage data, software, and expertise acquired by one department or faculty member to benefit the campus as a whole. GIS projects, like other digital data, can and should be archived within the library and included in institutional repository initiatives. These initiatives are gaining momentum within the library community.

- Technology support and leadership make a difference. Campuswide technical coordination and support by Information Technology staff are crucial for expanding GIS beyond single departments. For example, efficiencies are generated through consortium-wide software licenses and support from national organizations such as NITLE (*www.nitle.org*), from campuswide network support staff, from standardized hardware and software strategies, and from broad on- and off-campus training opportunities. Strategic maintenance of infrastructure is fundamental to enabling students and faculty to integrate GIS into innovative teaching and research projects.

- Strong administrative support matters. The successful growth and application of GIS across campus requires dedicated core faculty and committed departments and programs, but ultimately is not possible without strong support by the administration. By starting with a vision—in our case a campuswide strategic plan—the president and the dean of faculty were able to effectively mobilize external funds, support a new faculty line, and fund the expansion of GIS hardware, software, and human resources.

In conclusion, GIS is a particularly valuable tool to support environmental studies teaching, research, and outreach. At Colby, enthusiastic support of GIS by faculty, staff, students, and administrators is helping to integrate the technology into a spectrum of courses in the Environmental Studies Program and across the campus. The PES course demonstrates how GIS can support broader pedagogical goals, including supporting active inquiry-based learning, civic engagement, and interdisciplinary research and collaboration. We have been very satisfied with GIS and look forward to expanding its use in the future.

Acknowledgments

We gratefully acknowledge support we have received from ESRI for software, the National Science Foundation, the Andrew W. Mellon Foundation, President William Adams and the Board of Trustees of Colby College, staff of Colby's ITS Department, and the students enrolled in the Environmental Studies Program at Colby College. Multiple NITLE workshops were influential in the development of our GIS programs at Colby. Diana Sinton, Jennifer Lund, and an anonymous reviewer provided helpful advice on earlier drafts of this chapter.

Notes

1. Before taking this course, students are exposed to lake ecology concepts, study typical lake flora and fauna, learn field methodology, and are introduced to elementary GIS and GPS through the introductory ecology courses. These efforts help them to make connections among theory, field work, and geospatial analysis. A new GIS and remote sensing course was recently added to the environmental studies curriculum to provide more in-depth exposure to geospatial concepts and tools.

2. More than one hundred thousand depth points and their corresponding GPS coordinates were gathered using a high-end sonar coupled to a GPS receiver carried on one of the college's boats. Students cruised the perimeter of the lake and then made multiple transects of the lake on four different days, combining these efforts with water sampling efforts. Data points (simultaneous GPS and sonar readings using a Lowrance instrument) were taken every two seconds. Students generated a map of the completed transects so it was relatively easy to identify areas that required more coverage. Individual points were uploaded into ArcGIS software and converted into a raster grid where each pixel represents a unique depth.

3. A good-quality buffer was defined as covering the entire width of the shoreline associated with the property, having a mix of shrubs and trees, and having a depth of at least 50 feet (ideal is 75 feet). Most buffers lacked that kind of depth and some only covered a part of the shoreline.

4. Nine black-and-white aerial photographs (1":1000' scale) were used for the historic land-cover maps. Fourteen color aerial photographs with greater resolution (1":750' scale) were used for the current land-cover estimates. For both sets, photographs were rectified to digital orthophotos and control points on vector layers derived from thematic layers available through the Maine Office of GIS.

5. Soil was classified into nine classes based on data from the National Cooperative Soil Survey. The erosion potential factor (k) for each soil type was provided by the local Soil and Water Conservation District. This enabled the students to differentiate between soils that are less likely to erode (e.g., clay soils and soils commonly found in bogs and wetlands) and more likely to erode (e.g., sandy and silty soils). Slope was derived from a digital elevation model. Each land-cover class was weighted by erosion risk from low erosion potential (e.g., forests and wetlands) to high erosion potential (e.g., residential land, cleared land, and dirt roads). Proximity to the lake was included as a factor to account for the greater likelihood that soil or nutrients would enter the water from lands closer to the lake. A series of concentric 200-foot buffers were created from an initial 200-foot buffer close to the water extending out to the watershed boundary.

6. Colby participates in a consortium of Maine colleges and universities that provides educational discounts for ESRI software, including a campuswide license for ArcGIS and most ArcGIS extensions.

References

An Atlas of Maine. www.colby.edu/environ/courses/ES212/atlasofmaine.

Firmage, David H. and F. Russell Cole. 1999. Service Learning in Environmental Studies at Colby College. *Acting Locally: Concepts and Models for Service-Learning in Environmental Studies.* ed. H. Ward p.25–37. Washington, D.C.: American Association of Higher Education.

Nyhus, Philip J., F. Russell Cole, David H. Firmage, and Phoebe S. Lehmann. Enhancing Education through Research in the Environmental Studies Laboratory: Integrating GIS and Project-Based Learning at Colby College. *Council on Undergraduate Research Quarterly* 34 (September 2002).

About the authors

F. Russell Cole, Oak Professor of Biological Sciences, is a member of the Department of Biology at Colby College. He previously served as chair of the Department of Biology, the Division of Natural Sciences, and is currently director of the Environmental Studies Program. He is one of the core faculty who teach Problems in Environmental Science.

Susan W. Cole has been the science librarian at Colby College since 1978, where she provides library instruction for science classes, develops resource guides for individual classes and projects, and is spearheading Colby's library GIS program.

David H. Firmage, Clara C. Piper Professor of Environmental Studies, is a member of the Department of Biology and the Advisory Committee for the Environmental Studies Program at Colby College. He has served as chair of the Department of Biology, as director of the Environmental Studies Program, and as chair of the Interdisciplinary Studies and Natural Science Divisions. He is one of the core faculty who teach Problems in Environmental Science.

Philip J. Nyhus is assistant professor of Environmental Studies at Colby College, where he teaches courses in GIS and environmental policy. As a postdoctoral fellow funded by a National Science Foundation Award for the Integration of Research and Education to Colby College he was active in early efforts to develop GIS in the Problems in Environmental Science course.

Raymond B. Phillips is director of Information Technology Services and is an assistant professor in the Department of Biology. He and his staff have been responsible for implementation of GIS hardware and software infrastructure at Colby.

Daniel Tierney is a former teaching associate in the Biology Department at Colby College. He was GIS supervisor for the department and coordinated the GIS activities for the Problems in Environmental Science research course, independent study projects, and honors projects.

Edward H. Yeterian serves as vice president for Academic Affairs and dean of Faculty at Colby College. He has been a member of the Department of Psychology since 1978 and for many years has supported efforts to develop and expand GIS capacity at Colby.

Environmental assessments and watershed studies represent complex chemical problems, requiring integrated biological, geological, hydrologic, chemical, and demographic analyses (Harbor 1997). This chapter describes a project in which students at Denison University in Granville, Ohio, used GIS both as a visualization tool and as a means to further their interpretation of their lab-generated analytical results. GIS helped students assess a real-world spatial problem (lake sediment deposition as indicated by measuring concentrations and locations of lead in the sediments) and to visualize relationships between the factors that affect the patterns (soil types and land cover, for example).

Our collaboration began in 2000 with a chemistry senior honors project. Undergraduate student Shelie Miller wanted to apply her advanced analytical skills in a field experiment that would tie in with her environmental studies concentration. Brigitte Ramos, her chemistry advisor, and Karl Korfmacher, her environmental studies advisor, guided her work.

Brigitte and Karl were also interested in replicating this problem-solving approach in Brigitte's environmental chemistry course. As initially conceived, students in the chemistry course would collaborate with students taking Karl's Advanced GIS course to collect field data, create a GIS database and conduct spatial analyses, and use the results to make watershed management recommendations. These collaborations highlight the interdisciplinary links between the two sciences, expanding the learning experiences of both courses. Unfortunately, the collaborative pedagogical component of our project never reached its full potential, as we no longer teach at the same institution. Though we have begun our efforts anew at other schools, this chapter will focus first on Shelie's honors project, then Brigitte's subsequent effort to incorporate the project into her environmental chemistry class.

Chapter 14

Chemistry and environmental science: Investigating soil erosion and deposition in the lab and field

Karl Korfmacher, Brigitte Ramos, and Shelie Miller

Soil erosion and sedimentation are major worldwide environmental problems, degrading soil quality, altering hydrology, and impacting aquatic habitats and ecosystems (Pimentel et al. 1995; Thrush et al. 2004). However, measuring and correlating erosion and sedimentation rates is problematic, since not all eroded materials reach water bodies during a given storm event. Soil erosion models, such as the Universal Soil Loss Equation (USLE) and the Revised Universal Soil Loss Equation (RUSLE), have been used in GIS-based analyses (Engel 2003; Eastman 2003), but these models predict the amount of soil erosion within a field and not necessarily the amount that will be transported from the field and into a water body, such as our pond. Therefore, in situ sedimentation studies in ponds help us determine actual sediment deposition totals and derive sediment delivery ratios for watersheds (the amount of eroded material delivered to receiving waters compared to the calculated total amount of soil eroded in the watershed).

The focus of Shelie Miller's senior honors project was to learn whether the rates at which lead had been deposited in pond (benthic) sediments had changed over a thirty-year period (1970–2000). Because lead adheres to soil particles (particularly clay and silt), by measuring the concentrations of lead within a core of sediments, we could calculate the amount of soil that had been deposited into the pond since the point of known peak deposition. In small shallow ponds and reservoirs, such as our study site, much of the eroded material becomes trapped,[1] so we could also determine overall erosion rates through these methods.

Although lead is found naturally in the environment, these background levels are generally dwarfed by anthropogenic contributions from a variety of sources, such as paint, industrial materials and waste, ceramics, and fuel emissions. These rates were particularly high in the early 1970s, when the use of leaded gasoline peaked in the United States.[2] At that same time, lead was found in benthic (pond) sediments at its highest rates, measured in core samples in areas relatively free from industrial emissions (Callendar and Van Metre 1997; Van Metre and Mahler 2004). This suggests that lead can also serve as a temporal marker for sediment deposition studies in places such as our nonindustrialized watershed. Furthermore, because the watershed for our catchment pond is small (223 hectares), we assumed that rain would wash eroded soil into the pond relatively quickly. Therefore, 1970 was taken to represent the date of both peak lead atmospheric deposition (from vehicle exhaust) and peak lead benthic sediment deposition (from soil erosion and subsequent transport into the study pond).

Initial project: Shelie Miller's senior honors work

Early in her senior year, Shelie extracted 18 sediment cores, ranging in length from 5 to 115 centimeters from the pond at sites along transects that she and Karl had located with GPS receivers (figures 1 to 3). At each sample site, they also measured water depth so they could generate a current (year 2000) bathymetric surface of the pond (figure 4). Shelie and Brigitte used atomic absorption spectroscopy (AAS)[3] to calculate lead concentrations from the cores they had extracted. [For further details on all of these procedures, please refer to our *Journal of Chemical Education* article and supplemental laboratory documentation, "Implementation of a Geographic Information System in the Chemistry Laboratory: An Exercise in Integrating Environmental Analysis and Assessment." (Ramos et al. 2003)]

Since the top of each core sample represented the bottom of the pond in the year 2000, for each individual sample site Shelie could calculate the amount of sediment deposition since 1970 by measuring down to the point in each core at which the highest observed lead concentration was found.

Figure 1. Taking a sediment core in the delta area of the pond.
Photo by Brigitte Ramos. Used by permission.

Figure 2. Extracted core sample.
Photo by Karl Korfmacher. Used by permission.

If sediments were deposited uniformly throughout a pond, this process alone would be adequate to calculate the volume of sediment deposition (pond area multiplied by sediment depth). However, sediment deposition is not uniform, but rather nonlinear, since heavier sediments (such as sand) settle out more quickly than fine soil particles (like clay and silt). We know that this has happened in our study area over the last few decades since we see expansions of the pond's wetland and delta areas in 1999 aerial photographs (figure 3) that were absent in an earlier image (figure 5).

Because we knew that deposition rates, sediment depths, and peak concentrations would vary throughout the pond, we used GIS to interpolate raster surfaces from our sampled point data for both the current depth profile (figure 4) and the peak lead concentrations (figure 6). We chose a spline interpolation method and small (1 m²) cell sizes to produce a smooth surface and make it simpler to later calculate the sediment volumes.

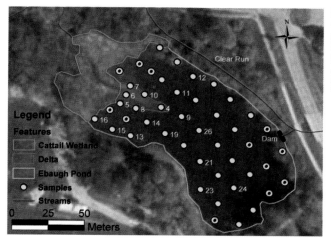

Figure 3. 1999 DOQQ with current (solid circles) and future (dotted circles) sample sites. Samples with numbers have been analyzed and were used in the interpolation analyses.

Reprinted with permission from the *Journal of Chemical Education* 80:1 (2003): 50–53; copyright 2003, Division of Chemical Education, Inc.

Figure 4. 2000 bathymetry image showing current pond depth. Red points indicate GPS locations of depth-only measurements. Not pictured are additional points showing depth data at the sediment sampling sites and points of zero depth along the pond shore. Image was generated using the spline interpolation algorithm with tension for all depth data. Notice the general trend of deposition continuing beyond the end of the delta feature, in the direction of the dam outlet.

Reprinted with permission from the *Journal of Chemical Education* 80:1 (2003): 50–53; copyright 2003, Division of Chemical Education, Inc.

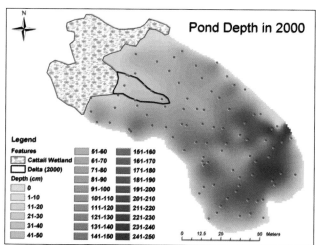

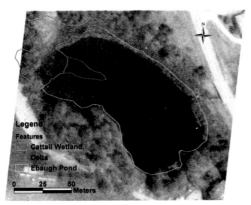

Figure 5. Georeferenced 1976 photo of the pond, showing no delta and a smaller wetlands area in the western portion of the pond. Lack of identifiable control points on the clipped image prevented a better fit to current pond features.

U.S. Department of Agriculture

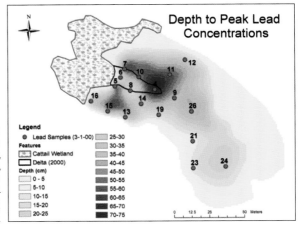

Figure 6. Interpolated depth to peak lead concentration image. As additional samples are processed, this image is expected to change, particularly in the eastern part of the pond, which is influenced by relatively few samples and does not account for the dam spillway.

Reprinted with permission from the *Journal of Chemical Education* 80:1 (2003): 50–53; copyright 2003, Division of Chemical Education, Inc.

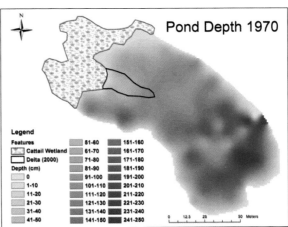

Figure 7. 1970 bathymetry image, created by adding depth to peak lead concentration to the 2000 bathymetry image. Notice the minimal delta development, compared to figure 4, and the relatively deep areas on either side of the current delta region. As the analysis of additional samples is completed, this image is expected to change to better represent the pond bottom in the early 1970s.

Reprinted with permission from the *Journal of Chemical Education* 80:1 (2003): 50–53; copyright 2003, Division of Chemical Education, Inc.

By using the difference in depth between the two surfaces (the depth in 2000 versus a surface of peak lead concentrations), we could calculate the volume of sediment deposition within the pond, since the peak lead concentration surface represents the bottom profile of the pond in the year 1970. The resulting values reflect an estimate of the pond's 1970 bathymetric surface (figure 7).[4]

These images allow us to see more clearly where and how the pond has changed as a result of thirty years of sediment deposition, and a 3D rendering enhanced this even further (figure 8). We also calculated sediment volume figures by converting the estimated sediment depth to meters, multiplying by the number of cells at each depth (since each cell is 1 meter^2), and then summing the totals. For our study pond, we calculated that 1,903 meters^3 of sediment has been deposited since 1970 (63 meters^3 per year), the majority of it settling out in the delta and wetland regions. Given that the watershed is 2,228,505 meters^2, as calculated by ArcView, this translates into a thirty-year soil loss depth of approximately 0.85 millimeter throughout the basin.

While the chemistry lab results seemed reasonable (peak lead concentrations ranged from 7.41 to 60.24 milligrams/kilograms with an average peak concentration of 27.11 milligrams/kilograms), the erosion and deposition estimates seemed surprisingly small, compared to other values seen in the literature for comparable watersheds. To understand why, Shelie integrated her knowledge of soil and earth science with chemistry to explain her results.

Figure 8. Three-dimensional views of 1970 and 2000 pond bottom interpolations generated with ArcScene. Formation of the delta region is clearly seen in the 2000 image, moving in the direction of the dam outlet.
Karl Korfmacher.

Our watershed contains a mix of land covers and land uses (figure 9). About 30 percent of the study area is cropland, a land use that is highly erosive. Moreover, several agricultural fields appear to be in the headwaters of the feeder creeks, suggesting that these areas might be the biggest contributors of sediment. Forests, however, protect soils from erosion, and in this watershed forests tend to be riparian (located near streams), which may cause them to act as sediment traps. When we inspected sections of the stream channels, we found extensive bank erosion in spots, particularly in areas adjacent to urban buildup.

The erosion susceptibility of a given soil type is generally determined by its texture and organic matter content, as well as where it is located on the landscape (the slope of the ground, and the length of that slope). Over half (56 percent) of the soils of the watershed are considered steep in terms of erosion potential (greater than 6 percent) (figure 10). The soil erodibility factor (or K-factor) is an estimate of the susceptibility of a given soil to lose particles through detachment and runoff from rainfall and surface flow during storm events. Soils in this watershed have relatively high K-factors (0.20 to 0.37), with 85 percent of the basin's soils assigned a value of 0.37.

Shelie concluded that although the watershed may be prone to erosion, much of the eroded material is still trapped within the watershed, assuming her estimates of sediment deposition were correct.

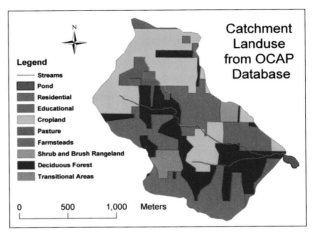

Figure 9. Land use and land cover within the pond drainage based on 1986 Ohio Capabilities Analysis Program (OCAP) data. While the feeder streams to the pond have significant forest cover along certain stretches, their headwaters are in agricultural fields that potentially produce heavy sediment loads.
Data from Ohio Department of Natural Resources and ESRI Data & Maps.

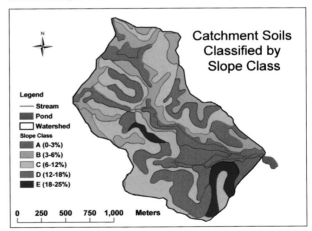

Figure 10. Soils of the pond watershed classified by slope type. Soils with a slope greater than 6 percent are usually considered at risk of erosion when modified by human activities, such as agriculture and development. While stream headwaters are within "A" slope soils, steeper slopes surrounding the headwater areas may contribute significant sediment loads.
Data from Natural Resource Conservation Service SSURGO and USGS.

Expansion to an environmental chemistry course

Shelie's project demonstrated that GIS could be used very effectively in conjunction with chemical analyses to address an environmental question. When Brigitte used the soil deposition project in her Environmental Chemistry course, students were presented with an overview of Shelie's work, including most of the images used in this chapter. The students were very enthusiastic about expanding the project and updating the GIS analysis, and they collected a number of additional sediment samples from the delta region, open water areas surrounding the delta region, and the wetlands. Shelie was even able to return to campus to participate in the data collection phase, showing the students how to use the piston core sampler and how to extract and prepare the sediment cores for analysis. The Environmental Chemistry students then processed the sediment cores and ran the lead analysis as detailed in Shelie's thesis methodology.

Wenzel and Austin (2001) and Dunnivant and Kettel (2002) make strong cases for linking unpredictable environmental issues to chemistry curricula in order to make the lab content and classroom experience relevant to the real world. Unfortunately, the students in the class conducted themselves poorly in the lab. When they began to analyze their results, it quickly became apparent that they were inadequately skilled to fully implement essential quality control practices. Brigitte observed many instances of sloppy lab technique and sample contamination (e.g., floating paper filters during the acid digestion procedure). She allowed the students to continue the experiment, leveraging the teachable moments that ensued as students discovered the shortcomings. Consequently, the data collected by this class was unsuitable for use in updating the GIS model. Although the Environmental Chemistry students were disappointed by this, they remained enthusiastic about the project, the goals of the exercise, and their participation in the experiment. In her next iteration of the project, Brigitte plans to spend more time reinforcing lab skills and discussing the importance of maintaining sample integrity and avoiding cross-contamination. In any case, the students were able to appreciate Shelie's work and understand her results in terms of erosion management for that watershed. For instance, they discussed alternative approaches to estimate erosion rates throughout the basin, including: (1) creating a GIS-based erosion model; (2) sampling within the growing wetland as well as the pond to see if that was originally part of the open water pond, hinted at in the 1976 aerial photo (figure 5); (3) collecting additional samples in the pond and delta region to verify the original estimates and the GIS model; and (4) further exploring the creek drainage.

Brigitte also feels the project has great relevance to core chemistry courses (she also teaches Analytical and Instrumental Analysis). By linking GIS mapping with data generated with atomic spectroscopy, the students could immediately visualize the spatial patterns of their data. Realistically, however, a chemistry instructor with little expertise in GIS would either have to devote much of a course to a project like this, or make a multicourse commitment to the use of GIS. Alternatively, the introduction of an existing, well-established GIS model (created in this case by Shelie) made the GIS integration manageable. Even though they were not able to complete their GIS analysis, the students could already envision the relevance of applying GIS to the study. They also appreciated being able to collect and analyze data for the GIS model without being overly burdened by having to learn all the details of the software at that time. Those interested in the software were encouraged to take GIS classes and/or minor in environmental studies.

Lessons learned in the field

Collaboration is critical to implementing cross-discipline curricular changes like the ones we began. None of us individually possessed the knowledge or skills necessary to make this change in a traditional classroom setting, yet each of us brought something valuable to the project. Our collaboration modeled the importance and appeal of interdisciplinary problem solving, particularly for undergraduate student theses, because the three of us were able to define and assign roles, skills, and tasks amongst our group. In teaching students and future scientists to successfully address these real-world environmental problems, we hope they learned most the importance of looking beyond the boundaries of their own discipline. This type of integrative research is invaluable preparation for future work situations. Particularly in environmentally related fields, collaboration between disciplines is essential.

Our own experience confirmed the documented benefits of student involvement in learning outside of labs and classrooms. Pursuing individual research projects or participating in class projects that are directly connected to real-world applications gives students a better understanding of actual problem-solving requirements and techniques. We recognized the value of providing students with an opportunity both to solve a problem and to make mistakes during the process. This type of preparation, and observation of faculty as they cross discipline boundaries, enhances the quality of the student's educational experience.

As we endeavored to incorporate GIS into Shelie's research and into the curriculum, we ourselves learned about the demands of adding real-life analyses to existing courses. Simply collecting, processing, recording, and storing the various field data takes considerable time and coordination (much more than we envisioned). Adding an advanced analysis (or two) to a field exercise can be rewarding, although the additional complexity may overwhelm students and faculty alike. Add to this the sometimes difficult GIS learning curve, and you have an unfortunately high threshold to curriculum reform.

GIS can be a powerful tool that integrates authentic environmental problems into the curriculum. The GIS maps allowed students to visually align their data, their lab results, the aerial photo, and their own observations of the current state of the pond. The calculations and spatial processing extended their understanding of the benthic environment, enabling students to visualize changes over the thirty-year study period. The 3D renderings let the students "see" the bottom of the pond as it was in 1970, understand the varying rates of sedimentation, and directly relate the computations to the physical soil loss across the watershed. Despite the significant effort involved in implementing the project, we and the students remain enthusiastic about the potential of using this sediment lead project as a central case study to integrate environmental chemistry and GIS courses.

Acknowledgments

We would like to thank the following people at Denison University for their help and support throughout the years we worked on this project: Tod Frolking, Julie Mulroy, and Chuck Sokolik. We would also like to thank the *Journal of Chemical Education* for allowing us to use figures from our earlier work with them.

Notes

1. As described in Van Metre, P. C. and B. J. Mahler (see references). Our study pond is dammed and our assumption is it functions as a small reservoir. The surface area to drainage area ratio is 0.005 (1:223 ha).

2. Callender and Van Metre (1997) estimate natural loadings of lead at 12 ktons/yr and anthropocentric loadings at 332 ktons/year in 1983. They also quote EPA figures from 1970, the year use of leaded gasoline peaked in the United States, showing automobiles emitting 182 kton/year of lead, with all other industrial sources accounting for only 37 kton/year. Tetraethyl lead ($[C2H5]4Pb$) was added to gasoline in the United States as an antiknocking agent (on average 2–3 grams of lead per gallon in 1973), but was gradually phased out over a twenty-five-year period following a 1973 EPA mandate. (EPA History-EPA Takes Final Step in Phaseout of Leaded Gasoline. www.epa.gov/history/topics/lead/02.htm [accessed Dec. 2004]).

3. Using a modified EPA standard procedure, Acid Digestion Procedure for ICP, Flame AA and Furnace AA Analyses (US EPA. 1998. EPA document, REAP-01.0, Exhibit D; sec. III, part C; D18-19). www.epa.gov/oamreg01/98-10007/ex-d.pdf.

4. These figures can then be rendered in 3D using a number of ESRI software extensions or products, including ArcView 3D Analyst (3.x), ArcGIS 3D Analyst or ArcScene (8.x), or ArcGlobe (9.x), to facilitate visual comparisons and highlight areas of change over time.

References

Callendar, E. and P. C. Van Metre. 1997. Reservoir Sediment Cores Show U.S. Lead Declines. *Environmental Science and Technology* 31(9): 424A–429A.

Dunnivant, F. M. and J. Kettel. 2002. An Environmental Chemistry Laboratory for the Determination of a Distribution Coefficient. *Journal of Chemical Education* 79(6): 715–17.

Eastman, J. R. 2003. IDRISI Kilimanjaro: Guide to GIS and Image Processing. Clark Labs, Worcester, MA.

Engel, B. 2003. Estimating Soil Erosion Using RUSLE (Revised Universal Soil Loss Equation) Using ArcView. abe.www.ecn.purdue.edu/~engelb/agen526/gisrusle/gisrusle.html (accessed December 2004).

Harbor, J. 1997. L-THIA—Assessing the Long Term Hydrologic Impact of Land Use Change: A Practical Approach. Ohio Environmental Protection Agency, Columbus, OH.

Pimentel, D., C. Harvey, P. Resosudarmo, K. Sinclair, D. Kurz, M. McNair, S. Crist, L. Shpritz, L. Fitton, R. Saffouri, R. Blair. 1995. Environmental and Economic Costs of Soil Erosion and Conservation Benefits. *Science* 267(5201): 1117–23.

Ramos, B., S. Miller, and K. Korfmacher. 2003. Implementation of a Geographic Information System in the Chemistry Laboratory: An Exercise in Integrating Environmental Analysis and Assessment. *Journal of Chemical Education* 80(1): 50–53.

Thrush, S. F., J. E. Hewitt, V. J. Cummings, J. I. Ellis, C. Hatton, A. Lohrer, A. Norkko. 2004. Muddy Waters: Elevating Sediment Input to Coastal and Estuarine Habitats. *Frontiers in Ecology and the Environment* 2(6): 299–306.

Van Metre, P. C. and B. J. Mahler. 2004. Contaminant Trends in Reservoir Sediment Cores as Records of Influent Stream Quality. *Environmental Science and Technology* 38(11): 2978–86.

Wang, X. 2001. Integrating Water-quality Management and Land-use Planning in a Watershed Context. *Journal of Environmental Management* 61:25–36.

Wetzel, T. J. and R. N. Austin. 2001. Environmental Chemistry in the Undergraduate Laboratory. *Environmental Science and Technology* 35(15): 326A–331A.

About the authors

Karl Korfmacher is an associate professor of environmental science at the Rochester Institute of Technology (Rochester, NY), where he teaches courses in GIS, environmental science, and environmental field studies. His research interests include mapping, monitoring, and assessing aquatic habitats (primarily wetlands); integrating GIS into the K–12 curriculum; and using GIS to help address urban environmental issues such as reclaiming brownfields and childhood lead poisoning.

Brigitte Ramos is an adjunct professor in the Department of Chemistry at Wilmington College in Wilmington, Ohio.

Shelie Miller recently completed her Ph.D. in civil and materials engineering at the Institute for Environmental Science and Policy at the University of Illinois at Chicago and is now an assistant professor in the environmental engineering and science department at Clemson University. Her interests include the sustainability of urban and industrial systems that incorporates the principles of pollution prevention, life cycle assessment, and green design.

Over the last two centuries, we have dramatically altered our land-use patterns throughout the world (Cunningham and Saigo 1997). One of the most significant changes has been the migration of people from rural areas into urban ones, and the conversion of those rural, previously agricultural places, to suburban and urban sprawl. In the United States, the urban population has increased from 5 percent to 75 percent and the rural population declined from 95 percent to 25 percent (ibid). Converting agricultural land to urban uses has profoundly impacted the natural environment at all spatial scales because of the unbreakable interrelationship between land, air, and water. Increased flooding, drinking water scarcity, degraded water quality, and rapid soil erosion are some local and regional concerns. At global scales, we are experiencing changes in precipitation patterns and temperatures that lead to glacier melting and sea level changes (Dunne and Leopod 1978; Schuler 1994; Bhaduri et al. 2000; Weng 2001; Booth 1991).

One way to represent the complexities involved in dynamic land-use changes is to model these processes with GIS (Guillen et al. 2001; Lim 2001; Markstrom et al. 2002). GIS' ability to organize and analyze spatial data and to generate results in an easily understandable, visual map format helps to convey research results and their implications, particularly effective for decision makers who may not have a scientific background. Undergraduate college students of today may become decision makers in the future, and I use GIS-based activities and models in several courses to help students understand the relationship between land use and hydrology. As a geologist, I use hydrological examples and case studies in my introductory GIS and remote sensing courses, and I will discuss those modeling exercises in this chapter.

Chapter 15

Geology: Long-term hydrologic impacts of land-use change

Suresh Muthukrishnan

Accurately measuring the effects of land-use changes on water quantity, water quality, and hydrologic flow patterns presents challenges. Water quality in streams and other water bodies is frequently affected by pollution, which we characterize as either point-source (PS) or nonpoint-source (NPS). Point-source pollution, from known and identifiable sources such as industrial and wastewater discharge, is controlled through stringent regulations and monitoring. On the other hand, nonpoint-source pollutants such as groundwater contamination and erosion are more difficult to monitor or regulate, particularly when they result from modified land-use patterns. Decision makers need a better understanding of how hydrologic processes and human actions lead to these pollution sources in order to address problems, develop management strategies, and implement policy changes. Models created through GIS can help provide this understanding.

GIS can seamlessly integrate multiple data layers efficiently, thereby simplifying hydrologic modeling and analyses. Hydrological models vary; some are used for engineering calculations that need to be as accurate as possible, and some are event-specific or short-term models that are used to study the impacts of a particular rainfall event. In my introductory GIS class, I use a model (long-term hydrologic impact assessment model or L-THIA) that falls into a third category and is useful for studying the long-term effects of land-use changes on hydrology and often used in comparative studies and policy development (Leitch and Harbor 1999; Grove et al. 2001).

The L-THIA model predicts relative changes in runoff and nonpoint-source pollutant load from one scenario to another (such as comparing a current pattern of land use to a proposed future one) (U.S. Dept. of Agriculture, Soil Conservation Service 1985). The model relates land use, soil, and rainfall through a simple relationship called the curve number method, developed and used by field scientists at the U. S. Department of Agriculture (Craig 1982). Curve numbers are proxies for runoff potential with different combinations of land-use, land-cover, and soil categories. A curve number links properties of soil, such as infiltration capacity, to the actual land use or land cover, which includes the imperviousness percentage in an area. Every combination of land use and soil type produces a unique curve number, and the model generates a raster grid whose cells contain these numbers, so students can examine their spatial distribution. The model uses long-term daily precipitation data, usually for thirty years or more, to compute the average annual runoff and also to determine antecedent moisture conditions (AMC), a measure of how much rainfall occurred in that location in a preceding number of days. AMC is important because continuous rainfall saturates the ground, leading to conditions in which all precipitation will run off the surface.

Logistics of integrating the hydrologic model

As part of my course, students learn to set up, run, and assess the results and implications of the L-THIA model. For one to two weeks we use the model by researching two different regional case studies. My objective is for students to learn how GIS can be applied to hydrologic research, so I spend lecture hours, in addition to one or two lab sessions, explaining the hydrologic principles used in the model and teaching students how to interpret the results. In my Remote Sensing of the Environment course, students also learn to process satellite images and generate land-use or land-cover data for different areas, data that become the input for the model that they also use in class.

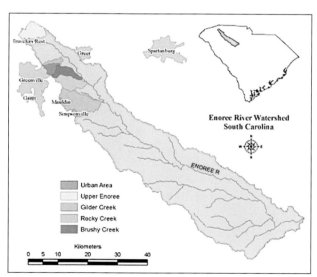

Figure 1. The location of the Upper Enoree, Brushy, and Rocky Creek watersheds within the larger Enoree River Basin.

Data from USGS Water-Resources Investigations Report 99-4015; E.P.A.; ESRI Data & Maps 2004.

Case study 1: Comparing watersheds with different levels of urbanization

Our first study area was the local Enoree River Basin, which encompasses the rapidly growing Greenville–Spartanburg area in upstate South Carolina (figure 1). Increased economic opportunities, availability of a good health-care system, moderate climate, and beautiful mountains have led to rapid development around the cities of Greenville and Spartanburg. To understand how a particular watershed may change with time, we needed land-use data for multiple years. Unfortunately, such data is not readily available for the study area, so we chose three small watersheds within the Enoree River Basin that are at different stages of urbanization to model hydrologic changes over time (figures 2) (Likens 1970).

Substituting space for time is common practice when studying geographical and geological processes over long time periods, and is acceptable for representing different land-use conditions when longitudinal data is not available (Craig 1982). Of the three watersheds selected, Upper Enoree is the least developed, with most of its land forested or in grass or pasture. Brushy Creek is the most developed, with over 60 percent of its area classified as high-density residential and commercial. Even though all three watersheds drain similarly sized areas (36 to 39 kilometers^2), variations in the proportions of each land use and each type of soil influence how much runoff is generated (table 1). When we compared the total volume of runoff produced by each watershed, using Upper Enoree as the reference point, students found that the more densely developed Rocky Creek and Brushy Creek generate much more runoff than the comparatively undeveloped Upper Enoree (269 percent and 329 percent respectively). We attribute most of this difference to the higher percentage of urban land uses. Although 35 to 50 percent of the watershed is either forested or grass or pasture, the higher percentage of urban land use causes most of the rainfall to become runoff that directly enters either the storm drainage network or the nearest stream. This excess runoff water causes frequent flash flooding in several parts of the watershed downstream of the most densely developed areas.

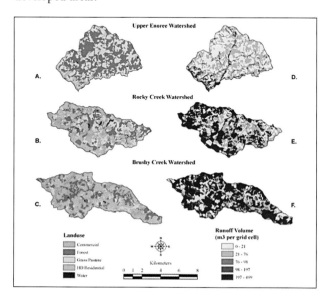

Figure 2. The spatial distribution of different land uses (A, B, and C) and model-predicted surface runoff volume (D, E, and F) for Upper Enoree, Rocky Creek, and Brushy Creek watersheds, respectively. In the runoff map, dark blue represents a higher volume of runoff, and light blue represents a lower volume, both per unit grid cell. Table 1 summarizes the runoff contribution by individual land uses.

Data from USGS Water-Resources Investigations Report 99–4015 and Suresh Muthukrishnan.

Table 1. The average annual runoff volume contributed by different land-use categories in cubic meters, and corresponding contributing area of each land use with respect to the total drainage area of Upper Enoree, Rocky Creek, and Brushy Creek watersheds.

		UPPER ENOREE	ROCKY CREEK	BRUSHY CREEK
Drainage area	(km²)	37.3	36	39.1
Grass/pasture	Runoff (m³)	477,143	277,992	144,500
	% of watershed	42.8	28.3	14.1
Forest	Runoff (m³)	358,861	138,541	116,929
	% of watershed	50.2	22.8	21.1
Residential	Runoff (m³)	457,567	2,387,288	4,413,842
	% of watershed	5.3	29.4	52.8
Commercial	Runoff (m³)	251,834	2,864,730	1,919,769
	% of watershed	1.6	19.2	12.1
Total runoff	(Million cubic meters)	1.55	5.67	6.6

Suresh Muthukrishnan.

Table 2. The model-predicted nonpoint-source pollutant loading values for the current and future land-use scenarios based on the average event mean concentration values. All units in kilograms.

LOAD (IN KILOGRAMS)	UPPER ENOREE	ROCKY CREEK	BRUSHY CREEK
Total nitrogen	1,832	8,503	10,850
Total phosphorus	374	2,288	3,147
Total suspended solids	35,286	258,449	290,348
Total dissolved solids	318,367	955,794	1,019,261
Lead	12	61	67
Copper	20	82	97
Zinc	90	713	709
Cadmium	1	5	5
Chromium	10	37	31
Nickel	8	58	67
Biological oxygen demand	18,943	127,450	157,898
Chemical oxygen demand	53,930	452,921	446,583

Suresh Muthukrishnan.

Students also investigate nonpoint-source pollution in these three watersheds and study the correlation between urban development and increasing nutrient and pollution loads (table 2). Increased nitrogen originates from lawn fertilizer, septic systems, and possibly the reduction in nutrient uptake by vegetation caused by the loss of forest and field (Likens et al. 1970). Increases are also noted for nickel, lead, and zinc, each likely resulting from high-density residential and commercial land uses and associated road networks. Even though the predictions consistently show much higher amounts of pollutants being generated from urban land uses, the amount of pollutants and nutrients that reach the streams and finally the oceans depends entirely on the types of land uses the runoff water passes through before it reaches the body of water.

As the runoff water travels through different land uses and stream conditions, a much more complex interaction between physical, chemical, and biological entities takes place that constantly modifies the water's chemistry. Hence, the relationships between land use and nonpoint sources of pollutants predicted by the model are complex, and a thorough examination of causality may be beyond the scope of an undergraduate project. This modeling exercise allows students not only to predict how different land-use conditions relate to the generation of pollutants, but also to understand the limitations of modeling in general. Running the model and knowing its limitations are both critical elements when planning patterns of development.

Case study 2: Comparing current and future land-use scenarios

We compare current and future land-use scenarios in the Gilder Creek watershed located southeast of Greenville and south of the Rocky Creek (figure 1). Employment opportunities and a strong public school system have begun to attract new residents to this eastern side of Greenville, where currently there is still much undeveloped land, unlike the neighboring Rocky and Brushy creeks areas. These factors portend future rapid development in Gilder Creek. For my students, this area represented not only a good site to model hydrology, but also an opportunity for them to understand how these types of modeling efforts can help local government make decisions regarding growth regulations.

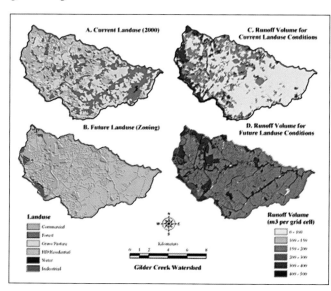

Figure 3. Spatial distribution of different land uses (A = present, B = future) and model-predicted surface runoff volume (C = present, D = future) for Gilder Creek Watershed for present and future conditions. On the runoff map, darker blue represents higher volume of runoff production and lighter blue represents lower volume of runoff production, both per unit grid cell. Table 3 summarizes the runoff contribution by individual land uses.

Data from USGS Water-Resources Investigations Report 99-4015 and Suresh Muthukrishnan.

We design two different scenarios, one corresponding to the current land-use conditions (figure 3) and the other corresponding to a future scenario with the entire watershed fully developed, as densely as county zoning allows. In class we run both scenarios through the L-THIA model and have found that if the future land-use changes are limited only by the current zoning regulations, then runoff is predicted to more than double (121 percent increase) in the future (table 3). In fact, with the exception of chromium and total dissolved solids, all other pollutants increase from current levels by 114 percent to 166 percent (table 4) because of forest and grass or pasture lands converting to urban areas. Since increases in runoff and pollutant loads result from specific land use and soil conditions, a community could minimize increases in runoff values and pollutant levels by focusing development in areas where soil has poor infiltration characteristics, while simultaneously restricting development in areas with more pervious soils or forested land cover. Students learn empirically that this "smart growth" method can benefit developers by making their property more valuable while at the same time preserving the hydrologic integrity of the land.

Table 3. The average annual runoff volume contributed by different land-use categories in cubic meters, and corresponding contributing area of each land use with respect to the total drainage area of Upper Enoree, Rocky Creek, and Brushy Creek watersheds.

		CURRENT LAND USE	FUTURE LAND USE
Drainage area	(km²)	82.4	82.4
Grass/pasture	Runoff (m³)	826,364	0
	% of watershed	35.4	0
Forest	Runoff (m³)	454,367	0
	% of watershed	27.6	0
Residential	Runoff (m³)	4,764,836	14,813,538
	% of watershed	25.1	77.5
Commercial	Runoff (m³)	4,056,467	7,466,950
	% of watershed	11.9	22.5
Total runoff	(Million cubic meters)	10.1	22.3

Suresh Muthukrishnan.

Table 4: Model-predicted nonpoint-source pollutant loading values for current land-use scenario and future land-use scenario based on average event mean concentration values.

LOAD (IN KILOGRAMS)	CURRENT	FUTURE	% CHANGE
Total nitrogen	15,157	36,911	144
Total phosphorus	4,063	10,806	166
Total suspended solids	428,132	1,024,911	139
Total dissolved solids	1,723,910	3,321,787	93
Lead	103	232	124
Copper	145	331	128
Zinc	1,140	2,571	126
Cadmium	9	19	114
Chromium	61	104	69
Nickel	97	234	142
Biological oxygen demand	218,071	543,647	149
Chemical oxygen demand	719,730	1,554,031	116

Suresh Muthukrishnan.

I have found these easy-to-use models beneficial in a classroom setting. The model relies on readily available data, rapidly generates visually informative results, and allows students to easily and iteratively manipulate input data to modify future conditions. Its input data includes land-use classifications, soil characteristics, and long-term precipitation values, information that is usually available for most parts of the United States and can be obtained directly from local, state, and federal agencies. With both case studies, I have students augment their computer-based modeling work with a written summary report, and when time permits, they present their modeled and mapped results and interpretations to the entire class. The written report demonstrates whether the students understand the principles behind the model or not, and their interpretation of the results indicates their ability to think critically, a skill that is necessary to connect the hydrologic process with the land use and soil conditions.

I have not been overly challenged in integrating these GIS-based hydrological models into my courses because of my own previous experience with the techniques. However, I encourage all faculty to experiment with these tools. The GIS-model interface is intuitive and comes with a step-by-step manual that walks users through different stages of preparing data, running the model, and interpreting the results using the tutorial dataset. Since GIS-ready land-use and soil data is already available for most parts of the country, many users need only delineate the area of interest and convert it to the format required by L-THIA.

Ways to incorporate GIS in the geosciences

Research objectives are changing in the geosciences and natural resource management, with a shift away from fixing a problem to instead first understanding the process that created the problem, and then trying to prevent it from reoccurring by taking appropriate measures. In turn, this affects the way colleges and universities educate students studying these disciplines. The geosciences field demands that people have both specific subject knowledge and skills in computer-based tools and technology, such as GIS and remote sensing. Today's geology majors need familiarity with GIS and modeling techniques for summer jobs and internships, and in their postcollege activities, whether they enter the workforce or graduate school. Departments must find ways for faculty to integrate technologies such as GIS and remote sensing into the core academic content.

Students enter my geology courses with varied backgrounds in GIS. Some may have already taken a whole term course in GIS, while others are only vaguely familiar with the concept of GIS and have limited or no exposure to GIS software. GIS can be applied to the geosciences in many ways since so many geologic processes have a spatial component. Thus it has become essential to add a geospatial component to process-based models, particularly those that require a wide variety of input data. The hydrological model discussed here was used in two courses, Remote Sensing of the Environment and an introductory GIS course, but these steps could also readily be incorporated into other classes that focus on hydrology, surficial processes, land-use planning, environmental sciences, or natural resources management:

- In courses with a GIS focus, a case study using this hydrologic model could provide a context for learning about raster data models and their significance to real-world environmental modeling. In addition, students could use the model to learn about spatial analysis techniques by doing a small project using spatial analysis and arithmetic operations on data layers.
- In a remote sensing class, hydrological modeling could be part of a small project spread over several labs. As a first step, students could use digital satellite data to process and perform classification techniques to generate land-use data in raster format. They could then use GIS to input the data required for a hydrological impact analysis. They could also modify the land-use data layer derived from satellite data to simulate growth and compare the corresponding changes in hydrological response.
- In a hydrology or environmental science class, a single lab or series of labs could be designed to teach hydrological principles and demonstrate how humans influence the hydrosphere. The runoff simulations that are run with this model could be used as a predictor for flood events and other related problems such as disease outbreak and drinking-water shortages.
- An environmental science class could also study the effects of population change and resulting growth in urban areas, measuring stress on the quality and quantity of existing water resources. Students could run this model and relate the runoff results to available water resources and consumption, population distribution, and regional climatic changes.

Any of these courses could expand to include community involvement, including sharing results with community groups or local government planners. Community planners would value access to data about the unregulated, less closely monitored nonpoint sources of pollution provided by this model. Students would benefit from contact with decision makers, the community would benefit from knowing the results of the study, and all parties might benefit from unanticipated outcomes from a community–university link.

Concluding thoughts

Increasing population and industrialization intensify the importance and complexity of hydrological interventions. Using a hydrologic GIS model across a range of geosciences courses introduces students to the subject of hydrology, introduces them to spatial analyses, and makes them aware of the critical need for planning and intervention in the early stages of urban development. Models and tools such as L-THIA and GIS support the development of systematic long-range and critical-thinking skills for our students, as well as encourage their understanding of, and appreciation for, smart-growth development.

Multiple human and geological factors contribute to hydrologic change in a watershed. Students benefit from using the GIS to see these factors visually organized and comprehensible. The ability to simulate this multifaceted system helps them understand the complexities of hydrologic systems and possible interventions.

Should my students enter the real-world politics behind land-use planning and development, they will be more effective advocates and decision makers because they are armed with a knowledge of the science behind the models. It may not be easy to persuade a developer or a township manager to use a GIS model to guide their decisions on watershed planning. Having had the experience of working with these case studies to analyze the development concerns of local communities, my students will be prepared and will understand the science, the process, the modeling tools, and the need to communicate clearly to all stakeholders.

Acknowledgments

I express my sincere thanks to Neil Gandy for processing the data and running the L-THIA model for the study areas presented here as part of his summer research project. That project was funded by an NSF-REU grant (EAR-0139135) through the River Basins Research Initiative program at Furman University. I also acknowledge the reviewers of this chapter for their valuable comments and suggestions.

References

Bhaduri, B., J. Harbor, B. Engel, and M. Grove. 2000. Assessing Watershed-Scale, Long-Term Hydrologic Impacts of Land Use Change Using a GIS-NPS Model. *Environmental Management* 26:643–58.

Booth, D. 1991. Urbanization and the natural drainage system—impacts, solutions and prognoses. *Northwest Environmental Journal* 7:93–118.

Craig, Richard G. 1982. The ergodic principle in erosional models. *Space and Time in Geomorphology*, Eds Colin E. Thorn, George Allen and Unwin Ltd.

Cunningham, W. P. and B. W. Saigo. 1997. *Environmental Science, A Global Concern*. 4th edition. New York: W. H. Company.

Dunne, T. and L. Leopold. 1978. *Water in Environmental Planning.* New York: W. H. Freeman and Company.

Grove, M., J. Harbor, B. Engel, and S. Muthukrishnan. 2001. Impacts of Urbanization on Surface Hydrology, Little Eagle Creek, Indiana, and Analysis of L-THIA Model Sensitivity to Data Resolution. *Physical Geography* 22:135–53.

Guillen, A., C. Meunier, X. Renaud, and P. Repusseau. 2001. New Internet tools to manage geological and geophysical data. *Computers & Geosciences* 27:563–75.

Leitch, C. and J. Harbor. 1999. Impacts of land use change on freshwater runoff into the near-coastal zone, Holetown watershed, Barbados: Comparisons of long-term to single-storm effects. *Journal of Soil and Water Conservation* 54:584–92.

Likens, G. E., F. H. Bormann, N. M. Johnson, D. W. Fisher, and R. S. Pierce. 1970. Effects of forest cutting and herbicide treatment on nutrient budgets in the Hubbard Brook watershed-ecosystem. *Ecological Monographs* 40(1): 23–47.

Lim, K. J. 2001. Enhanced NAPRA WWW spatial decision support system (sdss) with river water quality modeling capability. Unpublished PhD thesis. Purdue University, West Lafayette, Indiana.

Markstrom, S., G. McCabe, and O. David. 2002. Web-based distribution of geo-scientific models. *Computers & Geosciences* 28:577–81.

Schuler, T. 1994. The importance of imperviousness. *Watershed Protection Techniques* 1:100–111.

U.S. Department of Agriculture, Soil Conservation Service. 1985. National Engineering Handbook. Section 4, Hydrology. U.S. Department of Agriculture, Washington, DC.

Weng, Q. 2001. A Remote Sensing—GIS evaluation of Urban Expansion and its Impact on Surface Temperature in the Zhujiang Delta, China. *International Journal of Remote Sensing* 22(10): 1999–2014.

About the author

Suresh Muthukrishnan is assistant professor of earth and environmental sciences and director of the GIS and Remote Sensing Center at Furman University in Greenville, South Carolina. He teaches courses in GIS, remote sensing, natural hazards, and geomorphology. His research focuses on the use of remote sensing and GIS-based models to understand anthropogenic impacts on the biogeochemical cycling of nutrients in river systems and on water quality and surface runoff.

GIS and spatial thinking in the arts, humanities and languages
Introduction to chapters 16–17

Jennifer J. Lund

"The context in which you encounter something is going to shape how you understand it," states Pamela Fletcher, an art historian at Bowdoin College.[1] Fletcher uses GIS to map the emergence of the modern art gallery in nineteenth-century London. "Being hung in the foyer of a music hall, a high-priced gallery or a more formal exhibition society would have framed a painting in very different ways. This technology will allow me to ask questions in my research that couldn't happen otherwise—that doesn't happen often in my field."

Fletcher is studying geographic and social change through the decades in the city of London. Other researchers, describing their work in the following chapters, illustrate other ways to use the visual power of maps to organize, analyze, and represent research in the arts and humanities. With GIS, researchers create illustrative maps from their data to reveal stories about art, ideas, and artifacts, and the various people who made them, loved them, or lived them. Some maps show change across time, like Fletcher's study of art galleries; some show change across space, like the history of a town's prosperity written in its architecture.

A common goal of humanities, languages, and arts courses is for students to practice the work of the discipline, to use real tools, with real materials, in the authentic pursuit of knowledge. As educators, our challenge is to miniaturize the experience—scale down the reading, analysis, and commentary and fit it inside an undergraduate syllabus. Equally challenging, we seek meaningful work that students can accomplish with their novice-level skills. Too frequently an interesting topic is off-limits because students lack adequate knowledge and skills. Serious scholars of the discipline need a rich knowledge base and profound understanding to do the analyses *ex nihilo*. When we provide undergraduates with a prepared, self-contained GIS learning environment, they can produce illuminating maps despite having only a modest grasp of the theory and detail of the subject they are studying.

To make maps, plot observations, and convert numbers into images, we rely on complex and important technologies. Maps can position students on the edge of discovery, ready to discuss differences of context and culture, antecedents and influence, tastes and trends, audience and dissemination. Studying visual representations of interwoven factors, students view the information from a higher vista. Maps can provide a navigational metaphor for the sea of detail. Students are empowered to discuss the interplay of various social and artistic elements, analyzing patterns that emerge from a geographical perspective.

The chapters that follow describe classroom experiences of several innovative instructors in the arts and humanities who used GIS so their students could visually, rather than textually or numerically, reach an awareness of culture and context.

Robert Summerby-Murray's history class and Patrice Brodeur's religious studies course describe community service projects, emphasizing the importance of connecting with the community beyond the campus. Both authors describe the realistic hurdles faced by their students in the course of their research projects. Both describe how GIS served as a vehicle for presenting the results of their research. The religious studies example at Connecticut College pursued a goal common in undergraduate community service projects: to engage with "others" outside students' normal boundaries as a lesson in constructive citizenship. In both chapters instructors created a real-world context for the intellectual elements of their syllabi.

In the chapters about religion and music courses, students used GIS to visualize census data describing residents of their area of study. Students mapped their own observations of culture onto the maps to reveal the important truth that cultural phenomena exists in certain places, among certain people, because those people are in those places.

Three chapters describe the value of fieldwork assignments that engage students in the excitement of discovery. In history, religion, and one music project, GIS provided students with a structure in which to record and organize their data. GIS generated maps, presented data visually, and enabled students to analyze their findings by examining images.

Students thrive on the opportunity to seek and find, to build a personal connection with the material they study. Professor Travis Crosby, while teaching history at Wheaton College, let his students become acquainted with the residents of Victorian London. He used a map of the neighborhood populated with 1891 census data. Students could see the name of each resident along with his or her age, marital status, occupation, birthplace, and other characteristics. Crosby's objective was to engage students with the past. He describes the act of "doing history" as

> . . . rescuing the people and their lives and culture from the darkness of the unknown past. Our study brings them back to life for us, to know about them and their troubles and joys; to let them live on, in the memories and imagination of our time.[2]

224

His students enjoyed delving into the specifics of the census information, as if they were doing fieldwork in the past. They were excited by their observations and driven by their curiosity.

In each story in this section, students make discoveries about their own culture and others', seeing how their academic studies connect to people, ideas, and artifacts beyond the campus. In each case, maps help them make that context visible and available for analysis, discussion, and insight.

Notes

1. Pamela Fletcher. Web site accessed 11/1/2005 (www.bowdoin.edu/news/archives/1summerresearch/002387 .shtml).
2. Crosby, Travis. Professor emeritus. Wheaton College History Department. Personal communication. 2004.

A shared interest in GIS as a pedagogical tool catalyzed a collaboration between the French section of the Department of Modern Languages and Literatures and the Department of Sociology and Anthropology at Fairfield University. To prepare for this collaboration, foreign language faculty members attended summer seminars in GIS for four to five days. In addition, they read classical American sociological works, including essays from the Chicago School of Urban Studies.

Fairfield University's International Studies/Language Technology Initiative was funded in 1999 by the Charles E. Culpeper Foundation, The Rockefeller Brothers Fund, and the Archbold Charitable Trust. These initial grants supported a collaborative effort to bring GIS technology to the classroom.

Chapter 16

Foreign language and sociology: Exploring French society and culture

Joel Goldfield and Kurt Schlichting

Language, literature, sociology, anthropology, geography, and demography are natural allies for understanding a foreign country and its culture. Sociological information can illuminate the locations and characteristics of a culture's people and places. However, it is rare for language and literature courses to incorporate sociological elements at the undergraduate level. This chapter describes two projects that use GIS to study the population, culture, and landmarks of France. Both projects emerged from a collaboration between the Department of Modern Languages and Literatures and the Department of Sociology and Anthropology at Fairfield University.

In the first project, GIS provided a convenient user interface for interactive maps. Students are not normally attuned to the locations of people they read about in textbooks, newspaper articles, or novels, or of the characters they see in films. In the virtual world of GIS, students can rove the country, accessing multimedia illustrations of French cities and French people, building geographical awareness. The multimedia provides authentic sounds and images with the potential of engaging students more deeply than text alone. The GIS navigation tools allow self-directed exploration, which is more involving than passively watching images or video.

The second GIS project teamed students of intermediate French with sociology students to investigate census data and population patterns around Paris. They used the Internet to search for maps and census data, located and utilized information published by the French government, and used GIS to create maps to analyze their hypotheses.

These collaborations exemplify the goal of a program called Foreign Language study Across the Curriculum (FLAC). Beginning in the 1980s, private and federal grant initiatives encouraged schools in the United States to embed foreign language experiences in a wide range of courses.

Grants promoted language acquisition through the study of literature and culture within many academic disciplines. Language and cultural studies have meshed particularly well with the social sciences, as both pursue a deeper understanding of human beings and the world around them. Fairfield University's International Studies/Language Technology Initiative[1] is the first FLAC initiative combining GIS, sociology, and the teaching of French language and culture.

The initiative's efforts can be divided into four categories:

- Application of demographic and sociological theory to other fields
- Cross-disciplinary team building and problem solving
- Development of digitized maps for understanding culturally relevant, demographic data
- Development of new GIS-based applications for language acquisition, geographical knowledge, and the integration of relevant multimedia content

Faculty from the Modern Languages and Literatures, Sociology and Anthropology, and other departments created several projects in pursuit of these goals, two of which we describe here.

Accessing multimedia resources through a map

The first project, a multimedia map, was suggested when Fairfield's French faculty attended GIS training. One faculty member observed a strong parallel between the powerful navigation capability of GIS and a zooming technique pioneered stylistically by the romantic realist author Henri Beyle (Stendhal) in the first chapter of *Le Rouge et le Noir (The Red and the Black)* published in 1830 (Stendhal 1964, pp. 33–34). Multiple times Stendhal takes us from vast to intimate spaces, from the mountain range to the town and into the factory. Using GIS, users begin with a distant view and click to approach more closely, zooming in to see ever more detail. They finally reach a point where they can encounter a particular location in significant detail.

In their first GIS exercise, students began with a map of Europe. The instructor introduced them to tools for zooming, panning, and getting information about different elements of the map. On the map they navigated around the countries of Europe, with stars marking capitals. As they clicked on the stars, the names of cities and countries were revealed.

In their second exercise, students again began with a view of France and an introduction to the "hotlink" tool. Students roamed across France, navigating to various cities and clicking to see images, videos, and Web sites of landmarks and highlights of local culture.

Some video clips were familiar to students. For example, a short video of street athletes was taken (with permission) from their intermediate French unit on sport and the inner city in France (Thompson and Hirsch 2003). This engaging portrait of young French athletes provided a uniquely human look at life on the street near the La Glacière subway stop in the thirteenth district. Accessing the movie through a map presented these athletes in their geographical context. As the students zoomed in to the thirteenth *arrondissement*, they were developing an awareness of where the area was located in relation to Paris, and where Paris was located in relation to the rest of France as well as Europe.

Using the interactive map they created visual associations with places, and listened to pronunciations of geographical names. We had recorded the names and linked them to locations on the map.

Students could play and replay the audio as many times as they liked. Students using this map assimilated the pronunciations more accurately than in the past and were better able to recall the place-names and integrate them into conversation during classroom discussion. For example, when discussing writers and their native cities, we observed that students identified the cities on a map more easily and pronounced the French names more intelligibly and confidently.

Interactive maps have been used successfully as early as the third-semester level (the first semester of intermediate French) as well as in electives such as French Conversation and French Translation Workshop. Language and literature faculty can design the map and the assignments to adapt to a range of proficiency levels, making them useful for multiple courses. We added English subtitles and French captions that helped students understand the native French speaker narrating the video. Captions are available for different levels of mastery. We saw GIS maps improve students' ability to learn, remember, and talk about the locations of events and people discussed in their classes.

Learning French and sociology through collaboration

Sociological data such as census statistics contain a wealth of information about a country's population, level of education, ethnicity, national origin, median age, gender balance, professions, and other categories important to understanding a society. Maps help students visualize migration, location, and other sociological concepts much better than abstract statistics. We created a GIS project that focused on population density and created a cross-disciplinary assignment that teamed students from a French course with students from a sociology course. These teams investigated population patterns in Paris using a theory developed at the University of Chicago School of Urban Studies.

In the early twentieth century, a group of sociologists at the University of Chicago developed an important theoretical approach to the study of the city. Known as the Chicago School, Robert Park, Ernest Burgess, and others borrowed from the field of biological ecology to develop an ecological approach to studying the growth and structure of the burgeoning American city. Park and Burgess argued that cities grew organically. Burgess's concentric zone theory maintains that cities develop as a series of rings radiating outward from the downtown city center (Burgess 1925, pp. 47–62). Burgess argued that characteristic patterns of land use and population evolve within each successive concentric zone. For American cities, the outer zones are characterized by wealthier residents, less densely developed neighborhoods, and single-family, suburban-type housing. Implicit in Burgess's concentric zone theory is the premise that population density diminishes at a uniform rate as one moves from inner to outer zones.

The goal of the collaborative GIS project was for students to test Burgess's concentric zone theory on non-American cities, using the Paris metropolitan region as a trial case. Students addressed two questions. Is the French population surrounding Paris organized into a series of concentric zones that resemble Burgess's map of the city of Chicago in the 1910s (figure 1)? Does population density decrease uniformly as the distance from the center of Paris increases?

The students' first step was to search for the information needed for the project. Teams of French language students and sociology students used the Internet to search for digitized GIS maps of the Paris metropolitan region and recent census data. They canvassed a wide range of sites to determine whether such data was available, developing general research skills and cross-cultural awareness of French Web sites and government data.

We channeled students' attention toward the Institut Géographique National (IGN, the French national mapping agency) Web site to find digitized maps of the communes (small political units) in the Paris metropolitan region (*www.ign.fr*). Next, students went to the Web site of INSEE (Institut National de la Statistique et des Etudes Economiques), the French census agency, to look for the 1999 census population data at the commune level (*www.insee.fr*). Both Web sites required the French students to use higher-level language skills while the sociology students, trained in GIS, guided the search for the appropriate materials. The student teams determined which CD-ROM products they would order from IGN and INSEE, and determined how to place the orders, simulating international data acquisition that they may pursue in actuality in their future professional lives. Before placing the order, we delivered the happy news that the CDs had already been purchased.

With the CD-ROM from INSEE in hand, the teams were asked to identify data on population density for the communes in the Paris metropolitan region. Communes are among the smallest geographical units for which INSEE provides detailed social and demographical information. Students worked together to find the appropriate data on the CD.

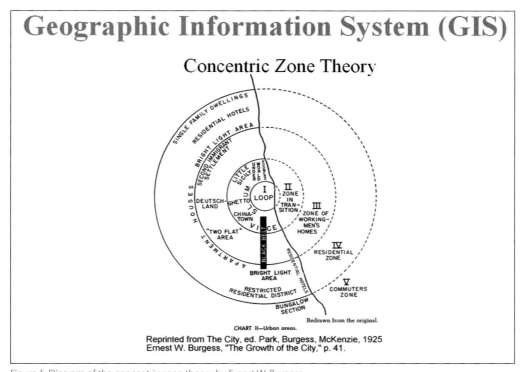

Figure 1. Diagram of the concentric zone theory by Ernest W. Burgess

Reprinted by permission from E. W. Burgess, "The Growth of the City: An Introduction to a Research Project" in *The City*, E. W. Burgess and R. D. McKenzie, eds., University of Chicago Press.

INSEE provides a series of helpful windows to guide the user and detailed descriptions of the data on the disk. The French students translated the "Territoires . . . Profils" that describes the *thèmes* or topics: population, age groups, household and family constituents, employment types, any necessary commuting, education, ethnicity, housing or living conditions, and more. Each team produced a spreadsheet with 1999 census data for all communes in Paris that included total population *(Population sans double comptes 1999)* and population density *(Densité de population [hab/km^2])* (figure 2).

The sociology students were already familiar with GIS and took the lead in the next phase, joining the spreadsheet data with the GIS map. This was straightforward because the French census bureau, INSEE, and the national mapping agency, IGN, use the same code to identify each commune. Figure 3 shows a map of the Paris region and Elancourt, one of the communes to the west of the city. In this map of census data, darker colors indicate denser population.

Students created multiple maps in their investigation of the Chicago School concentric zone theory. Their maps grouped communes into five levels of density, from very sparse (about 740 people per kilometer^2, or 3 people per acre) to very dense (more than 247,000 people per kilometer^2, or 900 per acre). Examining the series, they found that the pattern of population density in the Paris metropolitan region almost perfectly illustrated the Burgess concentric zone theory. Paris

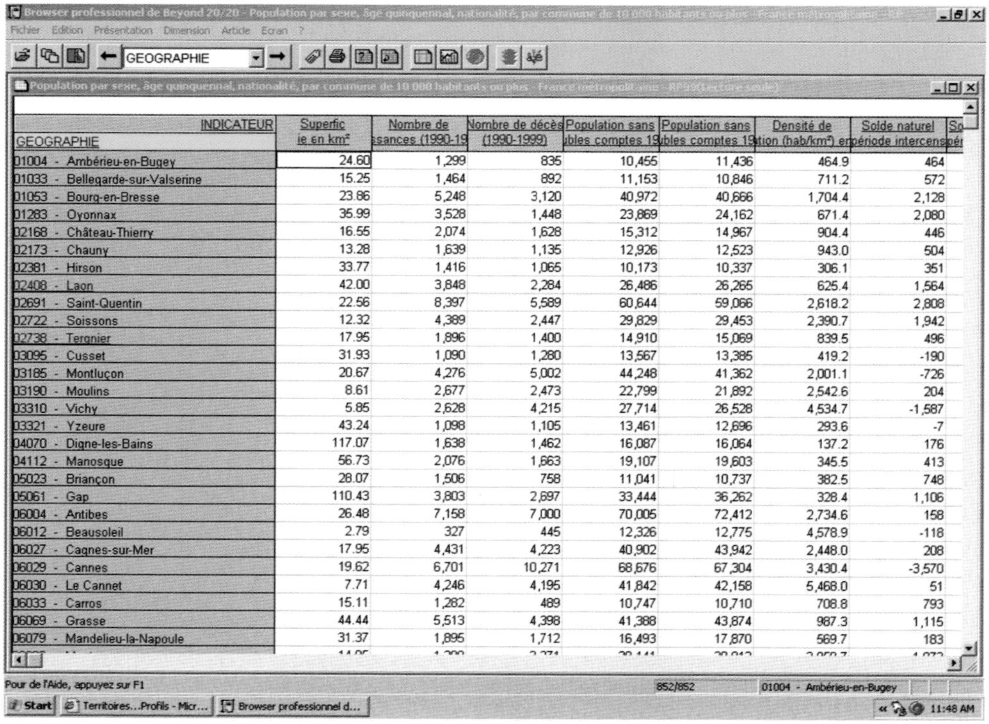

GEOGRAPHIE / INDICATEUR	Superficie en km²	Nombre de naissances (1990-19	Nombre de décès (1990-1999)	Population sans doubles comptes 19	Population sans doubles comptes 19	Densité de population (hab/km²) en	Solde naturel période intercens
01004 - Ambérieu-en-Bugey	24.60	1,299	835	10,455	11,436	464.9	464
01033 - Bellegarde-sur-Valserine	15.25	1,464	892	11,153	10,846	711.2	572
01053 - Bourg-en-Bresse	23.86	5,248	3,120	40,972	40,666	1,704.4	2,128
01283 - Oyonnax	35.99	3,528	1,448	23,869	24,162	671.4	2,080
02168 - Château-Thierry	16.55	2,074	1,628	15,312	14,967	904.4	446
02173 - Chauny	13.28	1,639	1,135	12,926	12,523	943.0	504
02381 - Hirson	33.77	1,416	1,065	10,173	10,337	306.1	351
02408 - Laon	42.00	3,848	2,284	26,486	26,265	625.4	1,564
02691 - Saint-Quentin	22.56	8,397	5,589	60,644	59,066	2,618.2	2,808
02722 - Soissons	12.32	4,389	2,447	29,829	29,453	2,390.7	1,942
02738 - Tergnier	17.95	1,896	1,400	14,910	15,069	839.5	496
03095 - Cusset	31.93	1,090	1,280	13,567	13,385	419.2	-190
03185 - Montluçon	20.67	4,276	5,002	44,248	41,362	2,001.1	-726
03190 - Moulins	8.61	2,677	2,473	22,799	21,892	2,542.6	204
03310 - Vichy	5.85	2,628	4,215	27,714	26,528	4,534.7	-1,587
03321 - Yzeure	43.24	1,098	1,105	13,461	12,696	293.6	-7
04070 - Digne-les-Bains	117.07	1,638	1,462	16,087	16,064	137.2	176
04112 - Manosque	56.73	2,076	1,663	19,107	19,603	345.5	413
05023 - Briançon	28.07	1,506	758	11,041	10,737	382.5	748
05061 - Gap	110.43	3,803	2,697	33,444	36,262	328.4	1,106
06004 - Antibes	26.48	7,158	7,000	70,005	72,412	2,734.6	158
06012 - Beausoleil	2.79	327	445	12,326	12,775	4,578.9	-118
06027 - Cagnes-sur-Mer	17.95	4,431	4,223	40,902	43,942	2,448.0	208
06029 - Cannes	19.62	6,701	10,271	68,676	67,304	3,430.4	-3,570
06030 - Le Cannet	7.71	4,246	4,195	41,842	42,158	5,468.0	51
06033 - Carros	15.11	1,282	489	10,747	10,710	708.8	793
06069 - Grasse	44.44	5,513	4,398	41,388	43,874	987.3	1,115
06079 - Mandelieu-la-Napoule	31.37	1,895	1,712	16,493	17,870	569.7	183

Figure 2. Spreadsheet of French census data.
Institut National de la Statistique et des Etudes Economiques.

arrondissements are comparatively densely developed. The communes immediately adjacent to the city of Paris are less so, and as one moves away from the core, population density declines uniformly. Beyond a 32-kilometer (20-mile) radius, few communes have high population density (figure 4).

To assess students' learning in this cross-disciplinary project, the sociology students had an additional assignment: to argue whether or not the GIS analysis supported Burgess's seminal work on the growth of the city. They compared their own maps with those from Burgess's first publication of the theory in 1925, and they found striking support for this aspect of the concentric zone theory in Paris. Students in the French class were asked to translate their new contextualized vocabulary in subsequent exams.

Connecting language and culture to sociology and geography

Students evaluated the cross-disciplinary collaboration as worthwhile because it provided successful opportunities to make unusual, yet useful, connections among sociology, GIS, and French language and culture.[2] The French students welcomed the GIS project as an opportunity to strengthen their language proficiency and content knowledge. A few sociology students had studied French previously and were able to reinforce their earlier language study. Both are important goals of our FLAC initiative.

Figure 3. Map of Paris, emphasizing the commune of Elancourt.

Data courtesy of Atlas des Franciliens, tome 3, 2002, © Iaurif-Insee; and Institut National de la Statistique et des Etudes Economiques.

Using numerical data, geographical data, and various related images to study the French language supports pedagogical theories of simultaneous left-brain and right-brain activities and multiple connections for improving foreign language acquisition (Hadley 2001, pp. 90–99; Met 1999). We discovered that fruitful discussions of deeper topics can occur when students are able to visualize the statistics and geography as people and places.

For example, we used the concentric zone maps (figures 3 and 4) in several courses at various levels to lead students to discuss historical examples of centralization. The class brought up centralization by French royalty and the derivative institutions such as the nationalized school system, army, policy, and taxation. When studying the nineteenth-century migration into Paris, students speculated on the behaviors of starving peasants seeking work and lodging near jobs in the city. They imagined models that would have produced the settlement patterns seen in the concentric zone maps. They reflected that when Louis XIV declared, "L'état, c'est moi," he foretold a migration that would put one quarter of the nation's population inside a 71-km (44-mile) radius. These discussions teach students to look at quantitative information like map legends and converse about density and proportion.

These kinds of connectionist activities, especially those involving quantitative and linguistic elements, are too seldom practiced in foreign language/culture classes, yet they are vital to students' ability to think and speak in real-life situations. Maps enable a multidimensional cognitive

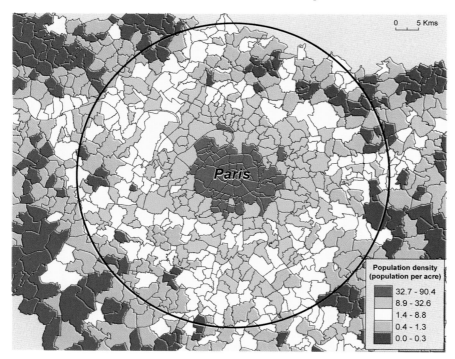

Figure 4. Map of the Paris metropolitan area showing rings of decreasing population density
Data courtesy of Atlas des Franciliens, tome 3, 2002, © Iaurif-Insee; and Institut National de la Statistique et des Etudes Economiques.

encoding, helping students develop a facility with the language essential to navigating the country-side, discussing politics, negotiating hotel rates, or debating restaurant prices.

The benefits of studying demographic maps accrue to all students, regardless of their proficiency with statistics or with GIS. Students in the humanities need not manipulate the statistical materials; they can access the material via maps produced by the instructor. Humanities instructors need not become GIS experts; they can use online mapping applications such as Google Earth, or enlist a colleague or a student proficient with GIS.

The collaborative nature of the French census project carried its own benefits. When students contributed their emerging language skills to a team, they demonstrated the value of their second language to themselves and their peers. The personal investment of using their knowledge in the service of research increased their commitment to the language. We observed students fulfilling these objectives of the FLAC movement.

The interactive multimedia map similarly engaged the students, increasing their facility with the language while making connections with the country, culture, and people. Students simulated moving into, out of, and across France, interactively controlling what they saw, and when. With this heightened degree of control, using their eyes, ears, hands, and curiosity, they learned more, remembered more, and were better able to contribute in classroom conversation. Students constructed a deeper understanding of the French culture by encountering information on multiple levels, in concrete and abstract terms, synthesizing diverse pieces of information from disparate disciplines.

Notes

1. FLAC has been successful in the liberal arts and professional schools alike, such as at the University of Rhode Island and the University of Connecticut. See, for example, press.ucsc.edu/archive/95-96/10-95/101695-National_conference.html. Also see the report on FLAC from the American Council on the Teaching of Foreign Languages at www.actfl.org/files/public/Fall1995LangAcrs.pdf. The best Web-based interdisciplinary resource making use of GIS for research and instruction and relating it to language and culture is the Tibetan and Himalayan Digital Library at the University of Virginia: iris.lib.virginia.edu/tibet/index.html.

2. See results reported on the Modern Languages, Literatures and Geographical Information Systems (GIS) Web page at www.faculty.fairfield.edu/jgoldfield/MLL-GISprojects.htm (Jan. 17, 2005).

References

Burgess, Ernest. 1925. *The Growth of the City: An Introduction to a Research Project.* Eds. Robert Park, Ernest Burgess and R. D. McKenzie. Chicago: University of Chicago Press.

Hadley, Alice O. 2001. *Teaching Language in Context.* Boston: Heinle & Heinle.

Met, Myriam. 1999. Making Connections. In *Foreign Language Standards: Linking Research, Theories, and Practices.* Eds. June K. Phillips and Robert M. Terry. Lincolnwood, Illinois: National Textbook Co.

Stendhal (Henri Beyle). 1964. *Le rouge et le noir [The Red and the Black].* Chronology and preface by Michel Crouzet. Paris: Garnier-Flammarion.

Thompson, Chantal P. and Bette G. Hirsch. 2003. Videotape to accompany *Ensuite: Cours intermédiaire de français.* 4th ed. Boston: McGraw-Hill.

About the authors

Joel Goldfield and **Kurt Schlichting** are faculty members at Fairfield University in Fairfield, Connecticut. Kurt Schlichting is chair and professor in the Department of Sociology and Anthropology. Joel Goldfield is chair of the Department of Modern Languages and Literatures (DMLL) at Fairfield University, director of the Charles E. Culpeper Language Resource Center, and associate professor of French, Foreign Language Pedagogy and Technology. Both share a deep interest in finding innovative ways to use technology in the service of their students' learning.

Robert Summerby-Murray's students are highly motivated by the prospect of influencing policy, and he saw in GIS an opportunity for Mount Allison University students to contribute to the local community. Summerby-Murray teaches various courses in practical and theoretical GIS, and the course described in this chapter was distinctive in that it responded to a community's need to better understand its own heritage architecture. The course required students to get their hands dirty with real-world data and produce policy recommendations for the town.

The class created a rich and detailed resource for local planners and preservationists that continues to inform the town. GIS supported a multifaceted teaching environment involving fieldwork, data collection, quality management, cartography, report writing, and public consultation. Equally rewarding, students said that the learning experience was constructive and enjoyable.

Chapter 17

Historical geography: Mapping our architectural heritage

Robert Summerby-Murray

This chapter describes how a course project using GIS and digital databases is helping a town preserve its heritage sites. The project's primary goal was to expose students to real-world data while applying theoretical concepts of urban heritage. The project asked students to construct research hypotheses about the built landscape and to test them using a database. Through readings and class discussion, students reached an important learning outcome by identifying, evaluating, and proposing remedies to counter the elite bias common to heritage preservation.

In contrast to some approaches to GIS that simply encourage unstructured exploration of a database, this project focused on theoretically derived research ideas such as testing for the presence of specific architectural forms and construction types, hypothesizing about changing settlement patterns, and critiquing the processes of heritage inventories in terms of social, political, and economic bias. Students created maps that enabled thematic analyses of historic building architecture and geographies of construction patterns between 1750 and 1914. Learning GIS was a derivative teaching objective, although it frequently dominated students' perceptions of the project.

I conceived of the project as a partnership with the local municipal government (through the Tantramar District Planning Commission), drawing in part on my involvement with municipal heritage issues. The town of Sackville, New Brunswick (a small town of approximately six thousand people), lacked a formal heritage management policy. As students developed their research questions and their formal written research papers, they heeded my imperative that project outcomes be relevant to policy making. The class was rewarded when policy makers implemented several of its recommendations.

Historical source material

I first developed the project in 1998 as the principal research experience of a one-semester, third-year undergraduate course in historical geography. The course comprised lecture, seminar, and laboratory components, and I introduced students to concepts of heritage landscapes, including their socially constructed and often controversial nature. The class spent time discussing the commoditization of historic landscapes for heritage tourism. They also encountered questions of historical accuracy, authenticity, and exclusion that often arise in the course of developing effective municipal heritage policy (Johnson 1996, Graham et al. 2000). With funding from internal university research grants, we hired a GIS technician from the local municipal planning authority and I negotiated with the planning authority, the municipality, and a provincial government department to gain access to data resources.

Students encountered several research challenges in the course of this project. The condition of the primary database posed the most significant problem. The 1989 inventory of Sackville historical buildings required significant updating before it could be useful. Students soon discovered additional challenges: questionable accuracy of the civic address database; the need to field-check architectural and property identification information; technical issues around integrating disparate databases; and mastering the appropriate tools to develop effective thematic maps.

Our primary historical source was an architectural database, the Canada Inventory of Historic Buildings (CIHB), developed by Parks Canada from a series of field surveys conducted across Canada beginning in the 1970s. The Sackville portion of the inventory was carried out in 1988 and 1989 and concentrated on residential, institutional, commercial, and industrial buildings constructed before 1914. It contained listings for 329 buildings, with data on dates of construction (both known and estimated), past and existing uses, and a series of detailed architectural characteristics including floor plans, foundation construction materials, exterior wall coverings, number of stories, window and door styles, roof types and styles, and other decorative features. In 1988 and 1989, data collectors recorded information on notecards with accompanying photographs, but the database was made available to the students in electronic format, with photocopies of the 1988 and 1989 photographs and coding sheets. With approximately seventy-five entries for each listed building, the database contained a wealth of information. Still, the database had not been applied to any serious discussion of historic building management in Sackville, and its use in other parts of Canada had also been limited.

The students offered two explanations for this. First, they discovered that some architectural historians had raised concerns about the accuracy of the original data collection, questioning both the training of the researchers and the quality of the survey. Second, Sackville's civic address system had changed several times, compromising the accuracy of the database and impairing students' ability to identify the listed buildings. While it was beyond the scope of this project to address the accuracy of the CIHB in its entirety, students were able to use the GIS to connect updated property information, first to the civic address databases and then to the Sackville portion of the CIHB. This connection was critical; it transformed the historic database into a relevant resource for local heritage planners.

Students experienced difficulties with other source materials, even those originating as municipal statistics. From the provincial government, we obtained accurate tax assessments, ownership information, mortgage information, and other georeferenced data. However, another civic address database, developed in 1998 for the introduction of an emergency 911 telephone system, contained numerous errors and inconsistencies. That database listed names of property owners and street addresses, but some were from an obsolete street numbering system. In many cases, the 1998 street addresses did not match the CIHB addresses from the previous decade. Students needed to field-check each database entry to ensure address consistency in the three data sources.

In the field

Small student teams were each assigned approximately 60 buildings from the 329 entries listed in the CIHB and provided with the old civic addresses as well as black-and-white photographs of the buildings taken in 1988 and 1989. Students used disposable cameras to document the updated appearance of buildings and recorded any obvious changes to architecture, use, or occupancy, such as conversion to a boarding house or a major renovation or addition.

Students had to deal with inconsistent (nonsequential and multiple) civic addresses. In some cases, buildings had held five different addresses since 1988. In addition, the black-and-white photographs taken in the 1988 survey were of poor quality and made positive identification of buildings difficult, particularly when supposedly unique architectural features were found to be reproduced on many buildings. While this proved frustrating, it also got students thinking about the spatial patterns of architecture in the study area. Another unforeseen difficulty was that the 1988 photographs were taken in midsummer and foliage often obscured parts of buildings. Students encountered a similar problem with their own 1998 photographs taken in mid-January: parts of buildings were obscured by snowdrifts!

The most frustrating field identification problem involved cases where buildings listed in the 1988–1989 inventory had been removed or destroyed. Confirming destruction was complicated further by the penchant of maritime Canadians for physically moving houses. Was the house demolished, or was the house on the other side of town that looked remarkably similar actually the house in question? The identification problem was so severe that by the end of the field component of the project, students had identified only 45 percent of buildings in the 1988 database. This represents a dismal performance for a database that was barely ten years old. Subsequent fieldwork increased this identification figure to 61 percent.

Although the project was time-consuming, students found the most rewarding element was field-checking to match the 1988–1989 entries with properties listed in the 1998 civic address database. Updating and revising the architectural and building-use data was likewise time intensive. The extensive effort challenged students to consider that any geographical information system is only as good as its database. Further, they realized that developing a usable database was not simply a mechanical laboratory exercise but involved creative group discussion, wherein they assessed the built environment in the field and raised hypotheses to confirm accurate identification of buildings.

Database development and GIS analysis

Entering data into the CIHB database posed few problems for students. Most were already familiar with basic spreadsheet operations and some had considerable experience working with databases and computer images. The GIS technician and I verified student work by reviewing data entries against early photographs and current tax assessment data, and following up on students' recommendations for the use of other sources. After each entry was verified, they scanned the new 1998 photographs that would be included on the CD-ROM of project output.

Students integrated the historical and contemporary information using ArcView software's ability to link nearly disparate databases. The property identification number (PID) from the provincial government's property information database served as a link to the civic address database and, by adding the same identification numbers to the CIHB architectural database, students linked all three databases. Their new architectural database contained updated civic address information (including specific property geocodes) and links to the database of digital images.

After linking the databases, students experimented with thematic queries. They created maps highlighting architectural features, construction dates, and other attributes of the historic buildings. It quickly became obvious that students were capable of much more than the original assignment of producing a single well-crafted map. Students rapidly acquired GIS skills sufficient to investigate their own questions, exploring novel permutations of various elements using the query and display capabilities in ArcView. Most groups chose to produce a series of maps to reinforce the conceptual point of their subsequent written analysis.

A new analytical tool for heritage planning

In the beginning, the students hypothesized that the CIHB reflected unstated assumptions about the designation "historic" and thus could be used to pose questions about the implicit values associated with heritage planning and related policies. They further developed their hypothesis through readings and class discussion. Most of the students' GIS analyses revolved around the nature of an elite historical landscape and how this was reflected in the built environment. They identified the shifting geographies of building construction over time, the use of specific building materials and architectural styles, and the representation of this landscape in historical databases and popular consciousness.

Initially their GIS queries and maps sought to illustrate the dynamics of settlement geography by sorting and displaying dates of construction. They began with the initial (but perhaps questionable) assumption that the listed historical buildings would be reasonably representative of overall housing construction. Students mapped buildings by decade, creating a series that suggested a spatial shift from north to south. They combined these decade-by-decade maps into one overview map (figure 1).

Middle Sackville, in the north of the study area, was a location of mills, tanneries, and light industry in the late eighteenth and early nineteenth centuries and thus contained the most examples from the database for the period 1750 to 1849. By the 1840s and 1850s, present-day Sackville had grown in the south in response to the bridge over the Tantramar River and had become an area of industrial, retail, commercial, and service activities. Indeed, of the buildings in the inventory constructed

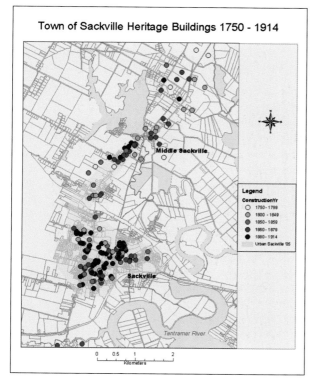

Town of Sackville Heritage Buildings 1750 - 1914

Legend

ConstructionYr
○ 1750 - 1799
○ 1800 - 1849
● 1850 - 1859
● 1860 - 1879
● 1880 - 1914
Urban Sackville '05

Middle Sackville

Sackville

Tantramar River

0   0.5   1   2
        Kilometers

Figure 1. Heritage buildings in Sackville, 1750–1914, dates of construction by combined decade groups. The oldest buildings (1750–1799) in the database are found in the north of the study area, in Middle Sackville. Throughout the nineteenth century, a progressive shift toward the south to present-day Sackville reflects the socioeconomic success and new buildings of the commercial and industrial elite. The basemap shows property boundaries.
Data courtesy of Canadian Inventory of Historic Building, Parks Canada, and Town of Sackville Civic Address Data Base.

241

between 1750 and 1799, 88 percent were built in Middle Sackville. By the 1850s, the number of listed buildings constructed in Middle Sackville had fallen to 37.5 percent, and by 1914 the pattern had reversed completely, with 90 percent of the listed buildings being constructed south of the study area.

As students looked more closely at building patterns prior to the 1870s, they found a higher concentration of listed buildings on particular streets, suggesting the evolution of high-end residential districts in the southern and eastern portions of present-day urban Sackville, associated with the expansion of Mount Allison University and the new homes of successful mercantile–industrial elites. Students correctly questioned whether the data accurately represented overall construction trends. Despite possible shortcomings in the sample, the students were successfully employing the CIHB database in the first systematic attempt to analyze late eighteenth- and nineteenth-century building patterns of this small industrial–commercial town. The hard-won victory highlighted for students the difficulties of using historical data that may be incomplete or have inherent bias.

They had similar questions when they analyzed the material history and architectural characteristics of the historic buildings. From their fieldwork and associated photography, students realized that the buildings of the laboring class were conspicuously absent. Larger, more expensive buildings and elite construction materials dominated the database, and they set out to document that finding. Using the GIS, students mapped the number of stories of the listed buildings (figure 2), showing both a dominance of larger, multistoried buildings in the CIHB (93 percent of the listed buildings) and a concentration of bigger buildings along the York Street–Bridge Street axis (see map inset), reinforcing a pattern seen in the construction-by-decade maps.

When they analyzed roof types, they saw a more complicated pattern (figure 3). The dominant styles in the survey involved medium and high gables (figure 4). Because these were used widely on both upper- and lower-status residential buildings, students could not conclusively relate this architectural detail with elite status. However, the geographic distribution of the higher gable style (with a concentration on the York Street–Bridge Street axis) approximated the clustered distribution of the more ornate and expensive styles such as the Mansard roof (figure 5), suggesting that these two styles were both indicators of higher-status construction.

Exterior wall construction and covering provided further architectural indicators of higher-status buildings. Students found that wooden framed buildings with clapboard exterior cladding dominated the listed buildings, reflecting a nineteenth-century New England vernacular tradition with

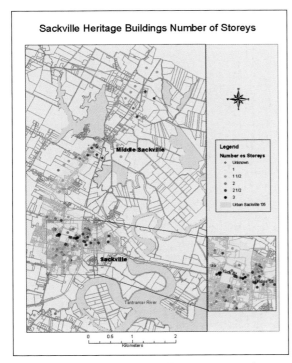

Figure 2. Heritage buildings in Sackville, 1750–1914, number of stories. Mapping the number of stories of listed heritage buildings reveals the dominance of multistory buildings in the database and indicates concentrations of the largest buildings in the late nineteenth-century elite residential district in the York Street–Bridge Street axis (see inset map).

Data courtesy of Canadian Inventory of Historic Building, Parks Canada, and Town of Sackville Civic Address Data Base.

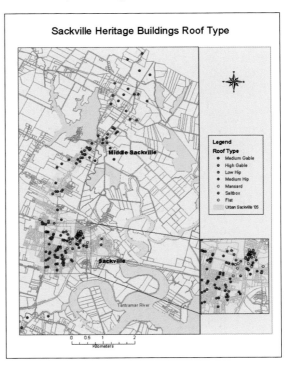

Figure 3. Heritage buildings in Sackville, 1750–1914, dominant roof types. While gable roofs dominate the listed buildings across the study area, this map indicates that more complicated forms (such as the Mansard roof) appear only in the south, in the mid-to-late nineteenth-century buildings of the higher-end residential district (see inset map).

Data courtesy of Canadian Inventory of Historic Building, Parks Canada, and Town of Sackville Civic Address Data Base.

Figure 4. An example of a decorated high-gable roof on a listed heritage building, Sackville (c. 1860).

Photo by Robert Summerby-Murray.

Figure 5. An example of a four-sided Mansard roof on a listed heritage building, Sackville (c. 1880).

Photo by Robert Summerby-Murray.

its emphasis on classical revival and carpenter gothic styles (figure 6). Although some architectural commentators have pointed to the use of stone exterior wall coverings as an indicator of higher-status residential construction, the Sackville survey suggests that high-quality cut stone was reserved almost exclusively for institutional buildings (such as those of Mount Allison University in the southern section of the town), reflecting the high costs of applying this material despite it being locally abundant.

Next, students mapped the presence of more expensive exterior wall claddings (pointed brick, cut stone, better-quality clapboard) on residential building facades as compared to the other walls. Their analysis revealed a distinctive geographic pattern: in areas settled first (Middle Sackville), strong concentrations of higher-quality materials were found on facades only (with lower-quality materials on other walls). Buildings in present-day southern Sackville, settled later, were dominated by structures where the same quality material was found on all exterior walls, usually the better-grade materials that had been limited to facades in previous decades and in earlier-settled locations (figure 7). Students hypothesized that increasing wealth associated with increasing industrialization reduced the motivation to economize on building materials.

Community outcomes: Contributions to heritage planning

From their analyses, students developed a series of recommendations pertinent to the local municipal planning environment. They had realized early in the process that the CIHB reflected an elite view of history, recording only buildings constructed prior to 1914 and dominated by large structures reflecting the social, economic, and political power of the mercantile and industrial elite. Students were excited by the possibility of influencing local policy. They carefully documented their findings, drawing upon spatial evidence of the socially constructed heritage landscape to question the CIHB's implicit definition of heritage. Their proposals challenged the exclusionary criteria of existing heritage documentation. They recommended that municipal planners and politicians consider a broader social view to manage the heritage landscape, one that could enhance tourism potential, residential and commercial environments, and local economic development.

Students were required to formulate concrete recommendations and to communicate them with written analyses to accompany their sets of maps. One student recommended that the town develop historical walking tours based on the building patterns in their maps (including some that involved specific architectural features or "houses of the 19th-century industrial and commercial elite"). Another student speculated on the possibility of creating a public-access GIS to generate individual walking tours based on users' preferences. Other students suggested various forms of historic building tourism. Some made specific recommendations for municipal legislation to protect buildings that were in danger of demolition or conversion. Some recommended establishing heritage districts based on prevalent architectural features. Significantly, a number of student papers called for heritage planners to look beyond the available "elite" database to engage the hidden historical geographies and heritage landscapes of labor, women, and the marginalized. Pointing out that Sackville is a small town with a long history of industrial employment, students persuasively argued that the lack of understanding of these heritage landscapes was a serious deficiency.

A small subset of students and I communicated directly to the town council. While the course project was not initially designed as an exercise in public participation GIS (PPGIS), it has assumed

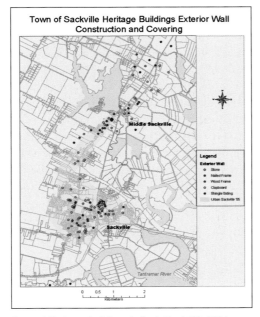

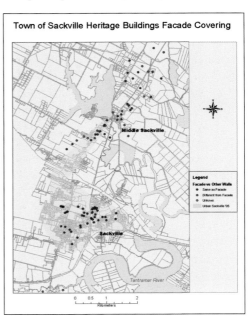

Figure 6. Heritage buildings in Sackville, 1750–1914, exterior wall construction and covering. Nailed or mortise-and-tenon wood framing provide the dominant exterior wall construction across the study area, usually associated with clapboard covering. Examples of shingle siding are found in Middle Sackville, while stone exterior walls are found on a small number of elite residential houses and institutional buildings in Sackville.

Data courtesy of Canadian Inventory of Historic Building, Parks Canada, and Town of Sackville Civic Address Data Base.

Figure 7. Heritage buildings in Sackville, 1750–1914, façade covering. For many of the older buildings (concentrated in Middle Sackville), the wall covering on the façade differed from the covering on the remaining walls of the building. Generally, the façade covering was a higher-quality material (such as a pointed cut stone or a dressed clapboard). The mid-to-late nineteenth-century heritage buildings in Sackville to the south of the study area show no evidence of this from the map, the façades being the same as the remaining exterior walls.

Data courtesy of Canadian Inventory of Historic Building, Parks Canada, and Town of Sackville Civic Address Data Base.

this role because of the town's ongoing utilization of the student analyses. This is consistent with the experience of other scholars (Ghose and Huxhold 2001; Bell and Reed 2004; Esnard et al. 2004; Warren 2004). This proves a useful example of how universities can partner with local communities. The Sackville town council responded to the project by strengthening its approach to heritage planning issues. It expanded its program of historic plaques and established a special committee of council (to which I was appointed) to advise on heritage landscapes.

The students delivered to the town analytical GIS maps, photographs, and recommendation papers on a CD-ROM. In late 2004, after much debate and public consultation, the town adopted a subset of a provincial statute providing for heritage policy and regulation. Sackville established heritage preservation areas, defined in part by the architectural and settlement pattern analysis from the student project.

Integrating historical geospatial data: Teaching GIS and heritage

The student evaluations I received focused on the technical aspects of working with databases and map display options and layouts. Taken alone, their evaluations would have reinforced the conventional wisdom that GIS was "too technical" to be used in historical geography. Despite our best intentions, GIS was not an invisible analytical tool. Barely half the students were geography majors and many others were sometimes overwhelmed by the technical elements of learning the software, even with the in-class guidance of the GIS technician. The technical nature of the project was reinforced by the numerous shortcomings of the data. Those difficulties provided valuable learning experiences, but they distracted students from their direct objective of thematic analysis. In contrast to the students' own perception of technical difficulty was the speed with which they learned to produce useful analytical maps, implicitly demonstrating that they were capable of overcoming the technical issues about which they complained. Overall, students expressed satisfaction that the project required them to explore real-world spatial data. They particularly valued their role in resolving the identification and accuracy issues of the CIHB database and engaging their problem-solving abilities and reasoning skills to effectively apply GIS to their field work.

Finally, despite student preoccupation with the GIS technologies, and despite the numerous unforeseen problems of using inherited databases, all students successfully engaged with substantive research themes within the study of heritage landscapes. GIS contributed powerfully to their understanding of both the past and the present, addressing conceptual issues of settlement dynamics, identifying geographic patterns of form and architectural type, and developing theoretically and socially informed positions on the management and planning of the heritage landscape.

Acknowledgments

The author acknowledges the work of students in Geography 3811 and is grateful for the financial and in-kind support of Mount Allison University, the Tantramar Planning District Commission, and the Heritage Branch of the Department of Economic Development, Tourism and Culture of the Province of New Brunswick. Research and technical assistance was provided by Stephen Clark, Laura Hope, and Heather Quinn. Additional cartographic work was completed by Christina Tardif.

Further reading

Bell, S. and Reed, M. 2004. Adapting to the Machine: integrating GIS into qualitative research. *Cartographica* 39(1): 55–66.

Bradbeer, J. 1996. Problem-based learning and fieldwork: a better method of preparation? *Journal of geography in higher education* 20(1): 11–18.

Dibiase, D. 1996. Rethinking laboratory education for an introductory course on geographic information. Cartographica 33(4): 61–72.

Ghose, R. and Huxhold, W. 2001. Role of local contextual factors in building public participation GIS: the Milwaukee experience. *Cartography and geographic information science* 28(3): 195–208.

Graham, B., G. Ashworth, and J. Tunbridge. 2000. *A geography of heritage: power, culture and economy.* London: Arnold.

Esnard, A., M. Gelobter. and X. Morales. 2004. Environmental justice, GIS, and pedagogy. *Cartographica* 38(3-4): 53–61.

Horne, M. 1993. The listing process in Scotland and the statutory protection of vernacular building types. *Town planning review* 64(4): 375–93.

Johnson, N. 1996. Where geography and history meet: heritage tourism and the big house in Ireland. *Annals of the Association of American Geographers* 86(3): 551–66.

Kemp, K., M. Goodchild, and R. Dodson. 1992. Teaching GIS in Geography. *The Professional Geographer* 44(2): 181–91.

Summerby-Murray, R. 1999. Heritage as elite construction: mapping patterns of historic buildings in small-town New Brunswick. *Proceedings of the New Zealand Geographical Society Conference.* eds. M. Roche, M. McKenna, and P. Hesp. Hamilton: New Zealand Geographical Society.

Summerby-Murray, R. 2001. Analysing heritage landscapes with historical GIS: contributions from problem-based inquiry and constructivist pedagogy. *Journal of geography in higher education* 25(1): 37–52.

Warren, S. 2004. The utopian potential of GIS. *Cartographica* 39(1): 5–16.

Wright, D., M. Goodchild, and J. Procter. 1997. GIS: tool or science: demystifying the persistent ambiguity of GIS as 'tool' versus 'science.' *Annals of the association of American geographers* 87(2): 346–62.

About the author

Robert Summerby-Murray is dean of social sciences and associate professor of geography at Mount Allison University in New Brunswick, Canada. His interests include urban historical geography, the contested nature of heritage, and the role of scholarly contributions to the development of public policy. He is active on municipal boards and committees to help develop policy that will protect urban heritage. His research on industrial heritage, cultural landscapes, and GIS education has been published in *The Canadian Geographer, The North American Geographer,* and *The Journal of Geography in Higher Education.*

The story of GIS and the Religions in New London seminar began when Patrice Brodeur attended a brief software demonstration by Beverly Chomiak of the Geology and Physics department at Connecticut College and began to imagine its potential for the academic study of religion. Like most faculty in religious studies, Brodeur had never heard of GIS. He was "no computer whiz," yet with the help of IT experts, had successfully experimented with the application of new technologies in previous courses. That experience persuaded him that technology is an invaluable tool for opening new ways of thinking about the academic study of religion, in both research and teaching.

Connecticut College is affiliated with the NITLE[1] consortium and had received funding for a program designed to attract faculty from the humanities and social sciences to learn about the many research, teaching, and learning opportunities available with GIS.[2] The information technology office was soliciting proposals from faculty, offering small grants and significant support to instructors interested in adding GIS to their courses. Brodeur drew up a brief description of the project described in this chapter and submitted a proposal. His grant proposal described an idea for integrating GIS into an advanced seminar that focused on the comparative ethnographic study of religion in this small New London town.

The excitement of having his proposal funded was tempered by Brodeur's anxiety over using the GIS software. He immediately sought collaboration with Chomiak, who held an appointment to support colleagues interested in GIS. Brodeur also received support from a librarian and an instructional technologist. Having no GIS knowledge himself, Brodeur relied upon Chomiak to help integrate GIS into the Religions in New London syllabus, to create maps with census data, and to teach the GIS portions of the course. Together they empowered their students to map field work results and relate them to U.S. Census data. Using GIS as a research tool, students illustrated for themselves how a population and its religious practices are intimately integrated.

Chapter 18

Religious studies: Exploring pluralism and diversity

Patrice C. Brodeur and Beverly A. Chomiak

The academic study of religion is by nature pluralistic—we study many faiths using multiple methods, and we look at both spiritual and physical worlds. This interdisciplinary nature encourages us to look at the connections between religions and the context of the believers, between religions and the physical world they inhabit. It also raises questions about social, economic, political, and historical factors that influence a religious community. In religious studies, students perceive how others' religious practices and communities are woven into the context of their own lives.

As we introduce our undergraduates to the methodologies of research in religious studies, we prepare them for a life of greater awareness and understanding. Often religious research focuses on interpersonal dialogue and humanist methodologies, yet data-oriented methodologies also have a place in our curricula. Just as students learn there are different ways to understand a religion, they should learn there are different ways to understand data, technology, and quantitative analysis. This chapter describes the experiences and responses of students when they were introduced to GIS and census data in a seminar entitled Religions in New London.

GIS meets Religions in New London

This seminar is described as "an ethnographic exploration of contemporary religious life in New London as a means to map its new religious diversity and explore the relationship between faith and American citizenship." New London, the small town where Connecticut College is located, has an interesting and varied history dating back to the prosperous days of the whaling industry.

Brodeur had taught this seminar previously in 1999 and 2001, when students gathered information on forty-four different religious communities of New London as part of a project called "The Pluralism Project at Connecticut College." This work was funded by various sources within Connecticut College and by the Harvard University Pluralism Project, which studies religious diversity in the United States.[3]

In 2002, GIS was not a central focus of the seminar, but rather a means to an end: a tool to enable students to understand the people in the surrounding city and make connections between their own research results and data collected by the government. We redesigned the final research project to include a significant GIS component. Many learning activities, assignments, and final assessments were modified to complement that project. This chapter describes only the activities directly related to GIS, focusing on the survey preparation, administration, and analysis.

The greatest modification to the syllabus was integrating community surveys and analysis of maps into the seminar's final project. However, we chose to introduce GIS and mapping concepts early in the course to demystify the technology. In the first week of class, Chomiak demonstrated GIS and had students map the religious communities in and around New London. To accomplish this, she conducted a two-hour session in the college's new GIS classroom.

For the initial mapping assignment, students divided the list of religious venues collected previously (churches, synagogues, temples, and so on) and transferred the data into a GIS table. Using the tables as input, the GIS software geocoded the street addresses and displayed them as points on the map. Each student struggled to learn the necessary GIS skills, but the exercise proved to be a useful initiation, successfully introducing basic GIS skills and concepts.

At the next class session, Chomiak shared the collective results of their efforts and students were excited to see the results so quickly. They immediately related the geographic cluster of religious venues with the historical growth of New London. Figure 1 shows the concentration of religious buildings in the old area of New London. This cluster represents both the legacy of the older religious communities that flourished in the city's prime, and also the newer but poorer and mostly immigrant communities that have migrated to the center of the city in the last few decades.

Next the class examined maps of U.S. Census data. Chomiak had created twelve maps of demographic characteristics including race, income, and ethnicity. She used the data corresponding to block groups, areas containing about 1,500 people.[4] Figure 2 shows the median income for each area and figure 3 shows what percentage of each block group in New London is Hispanic. The class looked carefully at these twelve maps and discussed spatial patterns—where maps shared common features and where they differed. Students identified the tight cluster of historic religious buildings as lingering evidence of a more prosperous era. They noted that this historic district has a fairly low median income, relative to other parts of New London, and that many of its residents claim Hispanic heritage. They discussed migratory responses to economic conditions, why people leave an area, and why people arrive. They reflected on how factors in the local and global economies interact to influence migration, how a city's population can change in response, and how that population change affects the nature and the identities of religious communities in a town.

The maps in figures 2 and 3 also prompted discussion of the wide range of demographic differences. Students saw a large range of median income and a large variation in the number of people

Religious Communities New London, CT

Legend
- ◎ Religious communities
- ⎯⎯ Streets
- ⌐ ⌐ Town Boundary
- ▢ Census tracts

0   0.25   0.5 Miles

Figure 1. Location of religious communities in New London, Connecticut.

Data collected by Patrice Brodeur. Additional data from ESRI Data & Maps 2004.

claiming Hispanic heritage. This exercise also taught students about census data: the types of information collected, what geographical areas are used, and what level of detail is available.

These early activities were useful because they established an expectation of success with the GIS software, while simultaneously introducing students to interdisciplinary thinking and moving fluidly between religious and social data. The census maps showed or implied income distributions, racial identity, and migration patterns. The instructors juxtaposed each of the twelve census maps with the map of religious institutions prepared by Chomiak. The class discussed their observations and the implications that arose when they viewed these images together.

The map of religious institutions represented their first general analysis of the religious communities of New London as a whole. Discussion focused student attention on the "ethnographic exploration" at the heart of this course and prompted conversations about the nature of the religious communities students would encounter in their ethnographic fieldwork. The students were excited by the visual learning opportunities in this first assignment, which helped create a positive attitude toward GIS. This outlook softened the impact of the more challenging GIS work to follow.

Data collectors

One objective of this seminar was to have students develop confidence in their abilities to collect and interpret their own research data. Their first task was to develop a survey to administer to the religious communities. The survey primarily addressed religious participation and demographics and was designed to complement the information available on the census maps. Because of the constitutional division of church and state, the U.S. Census Bureau is prohibited from collecting religious data. The students' surveys augmented the census data with important and illuminating information.

During this portion of the course, the class as a whole visited a different religious community every week. Brodeur concurrently assigned readings and ethnographic assignments not only to attune the students to the uniqueness of the religious community visited that week, but also to build the ethnographic skills they would need later to administer their own surveys.

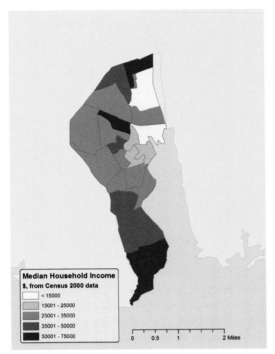

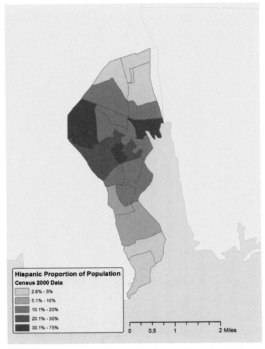

Figure 2. Income distribution from the 2000 U.S. Census.
Data from ESRI Data & Maps 2004; ESRI BIS.

Figure 3. Hispanic population distribution from the 2000 U.S. Census.
Data from ESRI Data & Maps 2004; ESRI BIS.

By the second half of the semester, students began to work more independently as data collectors. Working in pairs, they selected religious communities on which to focus and ascertained that their survey questions would be relevant to the members of their community. An important outcome of the seminar was for students to understand the process and the ethics of studying the "other." It was important for questions and possible responses to seem relevant to the survey respondents so the results would be meaningful to them. Students' research purpose was to create new knowledge, combining community data with census data, and to share that new knowledge with the communities they studied. The service-learning dimension of this seminar mandated that their efforts should benefit the community as well as themselves, providing a deeper understanding of respondents' own religious communities.

Another important goal was students' individual and collective learning about administering surveys. This was tremendously valuable both for its successes and failures. Seven out of the nine teams successfully administered the survey and were able to use GIS in their final papers. The two groups that were unable to gather survey data experienced negative reactions common in the world of survey research, probably the result of distrust and miscommunication. Another team paid for their success by committing an ethnographic transgression: one student was forced to become born-again in front of a congregation of over a hundred people! Fortunately, this person already considered himself a Christian and was not overly disturbed, but was glad his Jewish teammate was safely in the basement of the church, filling out surveys with Sunday school teachers. This experience was related to the class during one of the weekly debriefings. These debriefings benefited all students, precipitating a fascinating scrutiny of various types of ethnic blinders, challenging the students' worldviews.

Analysis and presentation

Conducting survey work has always been a component of this seminar, but the additional task of transferring the data into GIS maps was daunting for the students. Several teams did successfully map their own data. A few analyzed their own data in the context of the U.S. Census data. One team, Toni Ceci and Ronald LaRocca, examined the travel time between parishioners' homes and the Roman Catholic church St. Mary's Star of the Sea. Figure 4 recreates their map, showing circles at a distance of .5 miles, 1 mile, 2 miles, and 3 miles from the church. The map lists the percentage of English and Spanish speakers who live within those rings and how far they travel to reach St. Mary's. The map shows that more Spanish-speaking parishioners travel a shorter distance to church than most English-speaking parishioners. This was consistent with what the students had observed on the census maps, indicating that there was a high density of Hispanic residents living in the center of New London where St. Mary's was located.

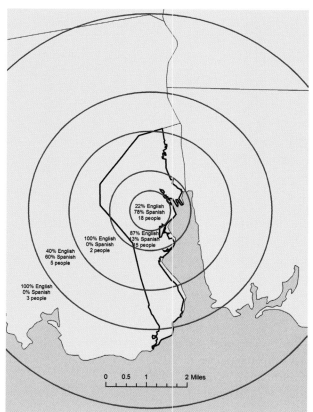

Another team, Laura Pollak and James Rogers, worked at the county level to map their survey data of the evangelical community of Faith Fellowship. Their data showed that most community members (89 percent) are U.S.-born and that the majority of community members live outside the city limits. This was consistent with what they had learned from the census data. Students were aware that in the United States, higher income levels generally correlate

Figure 4. Comparison of Hispanic- and English-speaking groups within one Roman Catholic community in terms of distance of travel between home and church.
Data from Patrice Brodeur and ESRI Data & Maps 2004; ESRI BIS.

with greater mobility; people with higher incomes are more likely to travel a long distance to their religious community of choice. Students were also aware that lower incomes generally correlate with newer immigrant populations. When they contrasted newer immigrant members and shorter travel times with older members and longer commutes, this pattern was confirmed.

Reflections on challenges

In hindsight, we realize we were somewhat overly ambitious in introducing so many software skills into this seminar. We have reflected on alternatives that would allow the course to run more smoothly, such as requiring students to have prior GIS experience or to master fewer GIS skills for the project. The current design was only feasible with the intensive team-teaching support of a GIS expert like Chomiak.

Students' chief complaint was the amount of time required to enter data and create maps. After spending many hours creating maps of their own data, they lacked the time to enhance their analysis with census data. The class also struggled with the unfamiliar demands of cartography. Their quantitative data was not displayed effectively, undermining the clarity of the final presentations. Following along with their verbal analysis was often difficult because the visual information on the map was ambiguous.

Although few groups achieved a substantial degree of integration between their own data and GIS maps of the 2000 U.S. Census, nevertheless the seven successful surveys did contribute important new information about each of their researched communities, strengthening the knowledge base of the Pluralism Project. Of the two teams that were blocked from administering surveys, one took the initiative to interview the other eight teams. Their second-order analysis, a study of the study, proved to be extremely useful to understanding the collective learning of the students in the seminar.

Students' major assessment for this seminar was not a traditional paper, but rather a live presentation in the context of an academic-style, half-day conference on campus entitled "Religions in New London: Contemporary Strengths and Challenges," a conference that actually took place a month after the semester ended. This scheduling decision was the result of negotiations with the hosting religious communities, exemplifying respect for the service-learning commitment and the administration's commitment to reach out to the town.

Students inserted their GIS maps into a Microsoft PowerPoint presentation and presented their findings to an audience of peers, faculty, and most unnerving, members and leaders of the very religious communities the students had researched. Unfortunately, only half of the religious communities availed themselves of this opportunity. However, representatives who did attend asked important questions, complementing the perspectives of the college participants. The conference brought a level of realism to this research and pedagogical enterprise that few courses manage to achieve in one semester.

Learning outcomes

This assignment integrated three principal goals: learning qualitative and quantitative ethnographic skills, developing and honing academic presentation skills, and mapping data. The students experienced firsthand the manifold challenges of doing academic research in the eminently inter-disciplinary field of religious studies. They bridged the traditional boundaries of data-centered social sciences and hermeneutically centered humanities in one seamless transdisciplinary educational experience.

By studying the diverse, though predominantly Christian, religious communities of New London, the students discovered the fascinating history of their immediate environment and gained a new understanding of how important those communities are to the fabric of contemporary life beyond the walls of their campus. They learned that it is impossible to study religions apart from other aspects of society; that aspects of society compartmentalized by academic departments are insepa-rably interwoven in the world outside the campus. Students also learned to value the way history shapes contemporary realities. They saw how local and transnational migratory patterns respond to a combination of local and global economic needs and affect the growth and decline of religious communities. Finally, they learned how to collect and process religiously sensitive information and make it visually accessible in maps.

Despite the challenges posed by GIS, there is no doubt that such a seminar is one of the most rewarding educational experiences for students of religions because it combines so many of the skills a religious studies course seeks to impart to them. Kristina Helb and Emily Serrell, who formed the team that interviewed their fellow seminar students, summarized many of the students' thoughts about this seminar in the following passage from their final public presentation:

> Judging by the results of our research, true cultural pluralism for our generation of students will be accomplished only when we can have enough positive and meaningful first hand experiences and interactions with individuals who are different from us. . . . I would never be able to grasp this from a book. . . . Everything is really based on how you are treated and how you treat them. . . . You may go in with a bias, but you will be more likely to walk out with reality if your experiences with the people there were positive ones. We cannot overemphasize the importance of actual human contact to the success of this project. Through this class, we have gained a feeling of ownership of New London. Just like those in their religious communities who were eager and proud to share their culture, now we will be eager to share about New London. We have also become more aware of how outsiders of the college must see us. Most importantly, we have made the first steps towards Pluralism by making meaningful human connections and have gained a feeling that, as a fellow student mentioned, "This was worth it."

Notes

1. NITLE is the National Institute for Technology in Liberal Education, funded by the Andrew W. Mellon Foundation and its affiliated members (www.nitle.org).
2. The grant for this project provided funds for faculty and staff education and small stipends for course enhancement. At Connecticut College, this grant was managed by Information Services, an organization that combines library and computer services.
3. The Pluralism Project, Committee on the Study of Religion, Harvard University. October 25, 2005. www.pluralism.org.
4. Block groups generally contain between 600 and 3,000 people, with an optimum size of 1,500 people. www.census.gov/geo/www/cob/bg_metadata.html.

About the authors

Patrice C. Brodeur holds a Canada Research Chair on Islam, Pluralism, and Globalization at the University of Montreal in the Faculty of Theology and the Science of Religions. Brodeur's research builds on and contributes to a unique Canadian tradition of interreligious dialogue as an approach to conflict prevention and peacebuilding. Born in Canada and educated also in Israel and Jordan, he obtained his PhD from Harvard University in 1999.

Beverly Chomiak is a member of the Department of Physics, Astronomy, and Geophysics at Connecticut College. Chomiak also maintains the GIS computer laboratory and spends much of her time there building a geodatabase of resources in the Connecticut College Arboretum. She helps other faculty in many disciplines integrate GIS into their curricula.

Matthew Allen, an ethnomusicologist at Wheaton College in Massachusetts, harbored a lifelong passion for maps. When he learned GIS was available on his campus, he quickly found a way to share his enthusiasm with his students by enhancing a music assignment with a mapping activity. He always asked his students to do field research, and now he asks them to collect spatial data about ethnic music venues and other cultural markers in neighborhoods where they could likely find Latin American music. At first the music majors were wary of technology, but as they got involved in the community and with the help of a GIS support person, they became motivated to use maps to document how ethnic heritage manifests in certain areas.

Allen's is one of several music-related projects that we describe in this chapter. Others include how GIS has been used to map a "Polka Belt" in the United States, understand subscription sales of music in eighteenth-century Europe, disseminate contemporary Christian music concerts, pinpoint Harlem venues of the 1940s, and map the concert career of a pianist freed by the Emancipation Proclamation. It is still rare to see music majors using and embracing GIS and mapping, but these projects are indicative of the possibilities.

Chapter 19

Musicology: Mapping music and musicians

Jennifer J. Lund

> In a way, [musicology] resembles archeology in that its art consists of piecing together
> with infinite patience and intuition every possible scrap of evidence. . .
> — Yehudi Menuhin (Stevens 1980)

Undergraduate instruction perpetually challenges instructors to seek opportunities for authentic and engaging work within the time constraints of a semester, and within the constraints of a novice's research skills. Collecting and combining minute pieces of information to create a whole is a task greatly aided by the tools of the information age, and GIS is a tool appropriate for the study of many aspects of musicology. Maps can help store, organize, and most importantly, see the emerging whole from the flurry of tiny pieces. This chapter describes a GIS project enhancing a course in Latin American music—one in which students enter their field observations into a GIS map. Then, because such examples are rare, we present a gallery of other applications of maps in musicology research, noting their potential for adoption or adaptation in undergraduate music courses.

GIS and Latin American music

In his Latin American music course, Matthew Allen has always included field work, sending students to hear the music and experience the culture they are studying (Allen 2004). Students in the course typically range across the full spectrum of cultural awareness, from those raised in Latin America to those completely unfamiliar with Latin American cultures and neighborhoods. Allen wants his students to be knowledgeable about the music and comfortable visiting venues where it is performed.

In the fall of 2004, Allen modified his fieldwork assignment to have students collect spatial data with their music and ethnographic observations. He asked his students to identify neighborhoods in nearby cities where they were likely to find Latin American music. Students visited those Boston-area neighborhoods in groups of two or three to search for ethnic music venues and other ethnic markers such as restaurants, hair salons, CD shops, churches, and community centers. Working in small groups increases students' power of observation and collective wisdom around personal conduct; it also augments their courage to find and communicate with the people in charge of the music-making event.

Students found a wide range of events, including clubs with published performance schedules, cafés with occasional music, church group performances, and festivals. Some venues were found via the Internet, but the assignment is structured for field observation and the most interesting venues were found by wandering—reading posters on lampposts and community bulletin boards. Their data sheets call for basic location information as well as country of affiliation, contact information, an estimate of audience size, and notes about musical format.

The assignment was timed to give students a weekend to visit the neighborhoods, then attend two evening lab times to enter field notes into a table in the GIS. A GIS support person and an instruction sheet provided assistance. The support person also created the basemap and provided a half-hour introduction to GIS to launch the assignment.[1] Although all the students were comfortable with communication technologies like word processing, e-mail, and instant messaging, not all were familiar with spreadsheets or working with data tables on a computer. Some students needed encouragement, but each group member took a turn entering his or her data points. For each venue, a point was placed on the map in its approximate location and students typed their observations and comments into a line of the attribute table. Figure 1 is a summary of data entered for a typical data point.

While it would have been most efficient for a student to simply type the address and observations into an Excel spreadsheet and mail it to the support person for geocoding[2], the instructor valued the act of placing the point on the map and having students enter their observations directly. The psychomotor participation establishes the students' personal connection to the dots on the map; it emphasizes their ownership of the map

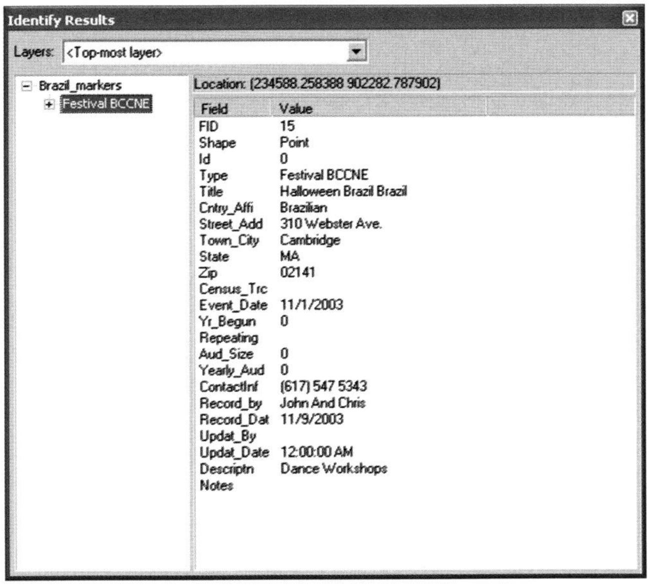

Figure 1. Data collected for a Brazilian Halloween party.
Data collected by Matthew Allen.

and their data, as well as their control over the technology. Many students expressed surprise and delight to see their own data displayed in such a polished manner, instantly integrated into the map. Most clicked on their points repeatedly, perhaps from enjoyment, perhaps from pride. After all groups finished entering their data, the support person collected all the points onto a single map and then placed the points more accurately, by street address.

The next class session met in the GIS lab, and students looked at the collective data, along with population maps of U.S. Census data. The census data, displayed as layers, showed what percentage of the population (in each census tract) claimed their ethnicity or birthplace as Brazilian, Cuban, Guatemalan, or from another Latin American country. Students could turn these layers on or off to see how the concentration of different ethnicities varied across eastern Massachusetts and Rhode Island.[3] Students also learned how to query the data, asking the GIS to highlight all points related to a specific country; for example, they learned to highlight all the Brazilian locations. From figure 2 they saw Brazilian points clustered together in neighborhoods, illustrating with their own observations how music is nested in culture.

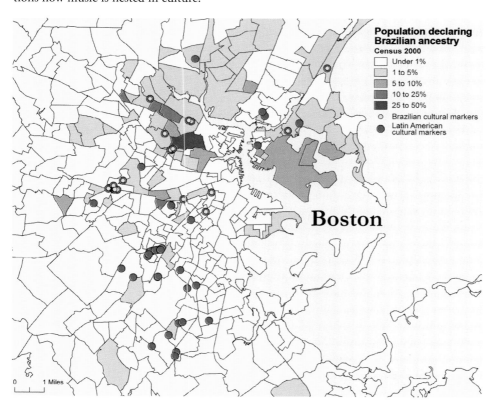

Figure 2. Brazilian cultural markers (yellow), and other Latin American cultural markers (red), on a map showing the percentage of the population declaring Brazilian ancestry in the Boston area using 2000 census data.
Data from ESRI Data & Maps 2004, U.S. Census Bureau, and Matthew Allen.

Figure 2 shows cultural markers found by students working in the Boston area. From this map, students can make the (unremarkable) geographical observation that some number of Brazilian events and storefronts were found in areas with high Brazilian heritage. Also from this map students can make the (more important) observation that they are capable of manipulating a data source to deliver information according to their interest. Because the GIS can present information in layers, students can pursue questions that interest them, comparing and contrasting information from various sources. They learn to ask a question of a dataset, get a response, and extract coherent information from a large body of minute details. The GIS can instantly create an image that visually summarizes multiple factors, which encourages students' interest in, and enthusiasm for, the data.

Using GIS in this simple way, to enhance a fieldwork assignment with geographic information, students collected their own evidence to document the interlinked manifestations of community and ethnic heritage. Working together, they constructed both a rich map and a shared awareness of the whole fabric of music, culture, and heritage, becoming ethnomusicologists themselves.

The assignment received a positive response from Allen's students, many of whom had expressed some resistance to using technology in this unfamiliar way. Students demonstrated a personal interest in their own data points, and demonstrated great interest in looking at the distributions of people and music venues related to the various ethnicities. Each individual wrote a short paper to complement the fieldwork and was given the opportunity to include maps in a final paper. Allen used the same exercise, with a similarly positive result, for his South Asian music course, and plans to continue having students add to this data collection each year, over time building a richer and more broadly useful map of ethnic music venues and cultural markers in the area.

Curricular benefits of using GIS in music courses

This data-oriented approach to musicology supports the basic tenets of a liberal arts education. What better way to embody the permeating importance of mindful and methodical research than in the study of musical expression and taste? The second part of this chapter very briefly describes research projects where maps illuminate music history and performance. Some highlight the roots of a phenomenon; some use geographic indicators to quantify success or change. Each map provides an opportunity to ask, "What does this tell us about the music? The musicians? The audience?"

These examples could be expanded or modified, at the instructor's discretion, as opportunities for undergraduate learning. Undergraduate research assistants or thesis students could assist with the data collection and preparation. They would learn perseverance and the important skills of problem solving and careful data collection, while learning the content and methodology of the discipline. Alternatively, maps depicting faculty research results could be embedded in coursework, enriching college courses by illustrating interwoven influences, and encouraging multidimensional thinking. GIS maps can serve as an interactive learning environment, in which students can view layers of a map independently, much like playing individual parts in a musical score, examining the interplay of factors, such as demographics and concert tours, or record sales and broadcast programming. Even brief exposures to such maps can support overall curricular goals by having students experience success with large datasets, plumbing the dark waters of data and extracting answers to their questions.

Mapping a musical diaspora

Populations migrate and take their music with them. Allen's Latin American music assignment introduced his students to that awareness. A recent study of the Polka Belt by Diana Sinton was conceptually similar but methodologically different (Sinton). The purported Polka Belt stretches from Connecticut to Denver. This study documented the statistically significant correlation between the "listening habits of a consistent dedicated audience" of polka aficionados and populations with ancestral ties to central and eastern Europe. Using Internet resources, Sinton identified 247 AM and FM radio stations in the United States playing more than two hours of polka music per week. She mapped those station locations and, estimating a conservative twenty-five-mile radius of signal reception, identified all counties within that broadcast range. She then mapped U.S. counties with the highest proportion of residents claiming central or eastern European ancestry, as reported to the Census Bureau. Even a quick glance at figure 3 suggests a visual correlation between polka-playing stations and Polish ancestry, which she corroborated with statistics to confirm the correlation. Sinton created similar maps for all ancestries she evaluated, essentially quantifying the influence of an ethnic heritage that may have persisted through multiple generations.

This model of study, applied to favorite musical genres, could generate great enthusiasm from both introductory and advanced students, with analysis ranging from "walking-around" data collection and visual inspection (as in Allen's course) to exhaustive research and statistical analyses (as in Sinton's study). Data summarizing ethnicity, age, education levels, and myriad other characteristics

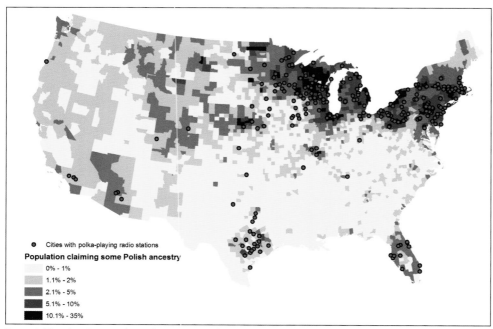

Figure 3. Map showing correlation between cities broadcasting polka music and proportion of residents claiming Polish ancestry in the U.S. Census, by county.

Data from Polka Bob, U.S. Census Bureau, and ESRI Data & Maps 2004.

is available from the U.S. Census Web site (U.S. Census Bureau). Other educators have used music to teach about patterns of migration, culture, and other social studies. ArtsBridge America, a federally funded program devoted to reviving arts education in the nation's public schools, developed such a curriculum for fifth and sixth graders (ArtsBridge America). Studying musical preference linked to locations on a map encourages students to think spatially and quantitatively on a topic of keen interest and personal significance.

Mapping dissemination

Popularity of a musical genre is conveniently measured by examining propagation. By looking at information about sales and performances, we uncover information about the audience, the musician and, by implication, the music. Tobias Plebüch, while an assistant professor of musicology at Stanford University, used GIS to illustrate his research into eighteenth-century subscription sales by C. P. E. Bach and D. G. Türk (Plebüch). Both composers were successful musical entrepreneurs who sold their scores to subscribers across much of western and central Europe. These subscription records provide valuable information about the tastes and demographics of musical audiences in eighteenth-century Europe. His data includes more than 10,000 purchases by more than 9,000 subscribers, representing more than 500 professions, and 57 geographical regions. The information includes name, gender, social status, and location of subscribers from 1752 to 1789.

Plebüch used GIS to map these records onto the subscribers' nation-states. Lists and charts illustrate variations in purchase patterns by nation and gender. Plebüch found that Türk sales were much larger overall, yet there were areas where Bach's work was much more popular. The map also generated information about the class divisions: 88 percent of subscribers were from non-aristocratic classes.

Plebüch created this map to support his own socioeconomic research into the social composition and spatial distribution of eighteen-century audiences. The GIS map made the data easily accessible, providing a way to slice data along multiple dimensions, to provide input to questions about, for example, discretionary spending according to gender.

Brent Hecht, a geography major at Macalaster College, undertook a study of the geography of Christian contemporary music (CCM), mapping the number of concerts played by top performers (figure 4) (Hecht). Hecht lacked the funds to purchase music industry data, so he collected concert tour information from the Internet. He then examined the CCM phenomenon by analyzing concert locations, church data, and county demographics. He notes, "As all good maps do, this map generated far more questions than it did answers. I now had a good idea of the regionality of the popularity of CCM in the United States, but I also had an intense desire to uncover the forces behind the patterns seen in the [concert location] map" (Ibid). He found that Christian denominational factors influenced the patterns more strongly than did the other variables he examined.

In contrast to Plebüch and Hecht, Roger W. Stump, a professor of geography and religious studies in New York, does not use location as a context for other measures of success, but rather uses location itself as a metric of success. Stump recounts the sites and experiences of the early innovators of bebop music, gathering with their peers at 4 A.M. at after-hours clubs in Harlem in

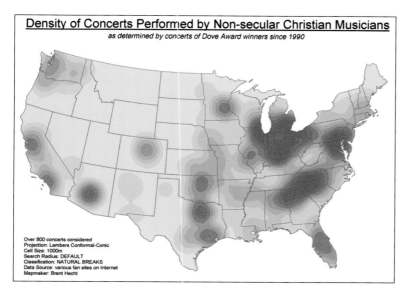

Figure 4. Density of concerts performed by nonsecular Christian musicians as determined by selected Dove Award winners from 1990 to 2003. Darker colors indicate higher density.

Map by Brent Hecht. Used by permission.

the 1940s. He quotes Dizzy Gillespie who was "enthused over the excitement of playing in Harlem" but recognized the importance of moving to the entertainment district extending across Manhattan (Stump 2003). As successful musicians began to play in clubs with swankier addresses, the music expanded out of its small circle in the predominant African-American neighborhood. Moving to 52nd Street, the hub of Manhattan night life, bebop was transformed into an accepted, mainstream art form, with larger audiences, better pay, and opportunities for recording contracts. Stump's study was mainly descriptive and focused on a small set of venues, and therefore was not the best fit with GIS. Research of this type, however, investigating more abundantly documented phenomena, could benefit from the organizational contributions of GIS.

Mapping musicians and musical groups

Another study informed by location was done by Wheaton College's Ann Sears, who documented the concert careers of nineteenth- and early twentieth-century pianists, particularly the artist known as Blind Tom (Sears 2005). Born a slave and freed by the Emancipation Proclamation in 1863, Thomas Greene Wiggins performed widely in the Americas and in Europe, under the controversial management of the white family from whom he was freed. Historical census data from IPUMS (Minnesota Population Center) will illuminate the racial and socioeconomic character of the cities where he played. Historical transportation maps will suggest the degree of convenience of his tours and how logistics changed over the years, as the western frontier of the United States opened to train and steamboat travel.

Sears chiefly used GIS to support her research into the management of musicians' careers. The project began when an undergraduate music major transformed a stack of photocopied clippings into a spreadsheet showing dates and places of performances. In the future, Sears would like to include data from other black musicians' careers. She would like to organize several decades of data to provide a rich resource for students writing theses on subjects ranging from audiences' changing tastes to racial hurdles faced by African-American artists of the period.

Three articles in an anthology focus on musicians' birthplaces. In one, John J. Flynn documents the change over time of principal singers in the New York Metropolitan Opera (Flynn 2003). In the 1900s, when the Met was relatively new, approximately three-quarters of the principal singers were European. Flynn's article details the transition from European to American singers, beginning when the company first appointed singers from the eastern United States and later from all across the country. By the end of the century well over half of the Met's singers were American-born, with the balance coming largely from Europe, with representation from Asia, Africa, and the rest of the Americas. In a second article, Blake Gumprecht profiles three successful country-rock singers raised in Lubbock, Texas, and cites their testimonies to how "the wind, the dust, and the environment" influenced their music (Gumprecht 2003). In a third article, George O. Carney, using data from three reference works, maps the birthplaces and northerly migration of several blues artists fleeing the south for Chicago's economic promise, from the 1920s to the 1960s (Carney 2003). Although these three studies, as reported, comprise too few data points for feature-rich GIS maps, they represent a promising approach for continued study looking at the influence of birthplace on artists' music, including what musicians bring to their performances and what draws the audiences to them.

Mapping the evolution of the guitar

Another model of study has been proposed by Matthew Allen (Allen 2005), mapping the various "technologies" behind the evolution of the modern-day guitar, using the "tour" metaphor described by Staley in this volume. GIS software has the ability to track movement through space and time and this capability would enable Allen to map, for example, the earliest historical references to frets, tuning pegs, or tuning systems in general. With adequate data, the chronology could be charted and compared to trade routes, diasporas, and intersecting cultures, enabling a student to extract a story line that could be broadened and deepened with the use of more traditional research sources.

Concluding thoughts

Over the last few decades, more musicologists have enriched their research with maps and geographical study. The practice of organizing and analyzing data with GIS continues to grow. The Information Age holds great potential for both research and teaching, allowing us to capture the seemingly infinite minute fragments of information in our fine net of knowing (Belenky et al. 1986).

This chapter has surveyed a number of models that could enhance music courses in a liberal arts college, teaching students about the subject at hand by illustrating the interrelated complexities of music, musicians, audiences, and the world in which they thrive. It can also advance the overarching goals of a college curriculum, teaching students to think spatially and logically. By putting authentic research within reach of undergraduates, we allow them to experience success collecting and using data, learning to value consistent research methods, and appreciating the robust data that results from structured inquiry. By having students collect and map their own data we demystify the process, helping them develop a healthy skepticism and to read published maps and digital data with a critical eye.

Many of our music majors, indeed many of our students overall, are in flight from numbers, technology, and structured study. Our new millennium demands a solid understanding and a minimal competency in all these areas, in all walks of life. Students are most likely to accept that message and instruction from faculty who share their interest in musical expression, but who also can apply the discipline of creating data to make maps that illuminate this shared passion.

Notes

1. Jenni Lund serves as an Instructional Technologist at Wheaton College, and is responsible for (and delighted to) help faculty with these processes.
2. Given a street address and town name, GIS software can place a point on a map, a process known as "geocoding."
3. Each town is assigned a color, from "very low" to "very high," divided according to "Natural Breaks." This method of assigning colors emphasizes the variation within an ethnicity. That is, even if the highest value is only 2%, variation will be visible with Natural Breaks.

References

Allen, Matthew. 2004. Music in Latin American Culture Course. Music Department, Wheaton College.

———. Associate professor of music. Wheaton College. Personal communication. 2005.

ArtsBridge America. Mapping the beat: a history and geography through music curriculum. University of California San Diego. artsbridge.ucsd.edu/Mappingthebeat/default.html.

Belenky, M. F., B. M. Clinchy, N. R. Goldberger, and J. M. Tarule. 1986. *Women's ways of knowing: the development of self, voice and mind.* New York: Basic Books.

Carney, George O. 2003.Urban blues: the sound of the windy city. Chap. 15, *The sounds of people and places.* Edited by George O. Carney. 4th ed. Oxford: Rowman & Littlefield Publishers, Inc.

Flynn, John J. 2003. Americans at the Met: the rise of the homegrown opera star in the twentieth century. Chap. 9, *The sounds of people and places.* Edited by George O. Carney. 4th ed. Oxford: Rowman & Littlefield Publishers, Inc.

Gumprecht, Blake. 2003. Lubbock on everything: the evocation of Place in the music of West Texas. Chap. 16, *The sounds of people and places.* Edited by George O. Carney. 4th ed. Oxford: Rowman & Littlefield Publishers, Inc.

Hecht, Brent. GIS and Christian rock. Unpublished manuscript.

Minnesota Population Center. IPUMS USA. University of Minnesota. usa.ipums.org/usa.

Plebüch, Tobias. Music subscriber statistics (C. P. E. Bach and D. G. Türk). Unpublished research.

Sears, Ann. 2005. GIS of Blind Tom's concerts. GIS project. Wheaton College, Norton, Massachusetts.

Sinton, Diana Stuart and William Huber. Mapping Polka and Its Ancestry in the United States. In press. *Journal of Geography.* Jacksonville, Ala. National Council of Geographic Education.

Stevens, Denis. 1980. *Musicology: a practical guide.* New York: Schirmer Books.

Stump, Roger W. 2003. Place and innovation in popular music: the bebop revolution in jazz. In *The sounds of people and places.* Ed. by George O. Carney. 4th ed. Oxford: Rowman & Littlefield Publishers, Inc.

U.S. Census Bureau. American FactFinder. factfinder.census.gov.

About the author

Jennifer J. Lund is the faculty technology liaison for the social sciences at Wheaton College in Norton, Massachusetts. She collaborates with the faculty to design class sessions and to teach students to use GIS for quantitative and spatial reasoning. Lund pioneered small-scale GIS assignments and integrated mapping into diverse courses across the college curriculum. Her passion for maps springs from her observation that students learn math from experience and observation as well as through logic and symbols. She earned a Master of Education degree from Harvard and has many years of experience in academic and corporate education.

Index

2D perspective, representing reality in, 23

AAC&U (Association of American Colleges and Universities), critical outcomes of higher education determined by, x
Aarseth, Espen J. (ergodic terminology), 45
AAS (atomic absorption spectroscopy), use in environmental science project, 202
ACS Course Delivery System, use in archaeology project, 158
active learning
    encouraging, 136
    promoting with GIS, 194
administrative support for GIS, importance of, 195
aerial photos, use in environmental studies project, 189–190
Africa, political and cultural variability in, 25–26
African Railroads map, 99
African reality, misrepresentation of West Africa, 113–114
African-American SES, role in Hurricane Katrina disaster, 21
agglomeration, economics of, 104
air pollution and acidic deposition, cognitive map of, 24
Alcock and Cherry, relationship to archaeology project, 156
Allen, Matthew
    Latin American music course taught by, 259–262
    mapping evolution of guitar, 266
AMC (antecedent moisture conditions), determining in geology project, 214
American cities, testing concentric zone theory on, 229–232
American FactFinder, description of, 107
American Indians, mapping concentrations in West, 85–87
antecedent moisture conditions (AMC), determining in geology project, 214
anthropology project (Centre College), 113–124
archaeology project (DePauw University), 155–164
    fall seminar, 158–161
    lessons learned, 164
    practicum, 161–164
architectural database, use in historical geography project, 238
ArcView map of federal lands in U.S., 101
arrondisements, dense development of, 231–232
arts courses, goal of, 223
Asian residents in eastern Massachusetts, mapping income of, 10
Asians, mapping concentrations on California coast, 85

Association of American Colleges and Universities (AAC&U), critical outcomes of higher education determined by, x
assumptions
    making from maps, 3–6
    questioning social assumptions, 8–11
atomic absorption spectroscopy (AAS), use in environmental science project, 202
Attleboro mobile home parks project, 93
Augusta, Maine, investigation of water quality in Togus Pond, 188
autism, mapping, 55–56
autism in Rhode Island, prevalence in school districts, 6–8

Bach, C.P.E., eighteenth-century subscription sales by, 24, 264
Baker v. Carr (one-person, one-vote rule), 131
basemaps, creating maps from, xv
bathymetric map, generation for environmental studies project, 188–189
bathymetry image for environmental science project map, 203
bebop music, innovators of, 265
Bednarz, Sarah Witham
    bios, 33
    and Diana Stuart Sinton, "About That G in GIS," 19–31
benthic sediments, presence of lead in, 201
Bertin, Jacques, use of maps as evidence by, 12
bicycle paths project (Ohio Wesleyan University), 64–68
Binger, Louis-Gustave (French explorer), 113
biology project (Kenyon College), 171–182
    analyzing forest structure, 172–173
    communicating results, 176–177
    mapping islands and population ecology with GPS and GIS, 174–175
    spatial analysis, modeling, and fieldwork for wetland restoration, 177–179
    transforming spatial data, 175–177
    working in field and landscape planning, 179–181
bioregionalism
    goals of, 64
    roots of community mapping in, 63
black residents in eastern Massachusetts, mapping income of, 8–10
Blackburn, John, 79
    bio, 138
    "Political Science: Redistricting for Justice and Power," 129–137
blacks, mapping concentrations in Southeast, 85

Blind Tom, concert career of, 265
Bloom's Taxonomy of Learning, 20
BLS (Bureau of Labor Statistics) unemployment map, description of, 108
Booker, James
    bio, 110
    "Economics: Exploring Spatial Patterns from the Global to the Local," 97–108
Booth, Charles, map of Whitechapel district (1891), 12
Boston, musicology project conducted in, 260
Boston area, residents with Brazilian ancestry in, 261–262
Boston census data, 2000, 50
Boston income disparity map, 54
Boud and Feletti (problem-based instruction), 143
Bowdoin Scientific Station, location of, 174
Boylan, Miles on GIS in economics, 106
Brazilian ancestry in Boston area map, 261
Brazilian Halloween party data collection (summary), 260
Brodeur, Patrice C.
    bio, 256
    "Religious Studies: Exploring Pluralism and Diversity," 250–255
Brushy Creek watershed, runoff for, 215–216, 218
buffering, using in Habitat for Humanity project, 148
buildings
    analyzing in historical geography project, 241–242
    documenting in historical geography project, 239–240
Bureau of Labor Statistics (BLS) unemployment map, description of, 108
Burgess, Ernest (Chicago School), 229

C4 (Chester Consortium for a Creative Community), collaboration with Swarthmore College, 70–71
Caine and Caine (Making Connections: Teaching and the Human Brain), 11
calculator, considering GIS user interface as, 56–57
campuswide GIS, lessons learned from implementation of, 194–196
Canada Inventory of Historic Buildings (CIHB) database, use of, 238
Carleton College buckthorn identification and eradication project, 66
Carney, George O., documentation of musicians' birthplaces by, 266
cartography, relational cartography, 44
Carver, Jonathan (New Map of North America, 1781), 5–6
case studies, approach toward, ix, x
Catchment Landuse from OCAP Database, 206
Catchment Soils Classified by Slope Class, 206
categorization
    mapping drive toward, 3–6
    role in problem solving, 23–27
Cayuga Lake watershed basin project, 104
CCM (Christian contemporary music), geography of, 264
census and decennial reapportionment, 130–131
census data from Victorian London, 11–13, 224–225

centralization, mapping historical examples of, 233
Centre College anthropology project, 79, 113–124
    reflections and lessons learned, 123–124
    spatial analysis of Kilimi region, 116–123
CGMA (Collaboratory for GIS and Mediterranean Archaeology)
    fall seminar, 158–161
    lessons learned, 164
    objectives of, 155–156
    practicum, 161–164
    relationship to MAGIS, 157
CGMA faculty and students in breakout sessions, 161
CGMA students using laser transit, 160
chemistry curricula, linking environmental issues to, 207
chemistry and environmental science project (Denison University), 201–208
    expansion to Environmental Chemistry course, 207
    lead deposits in pond sediments, 202–206
    lessons learned in field, 208
Cherry and Alcock, relationship to archaeology project, 156
Chester Consortium for a Creative Community (C4), collaboration with Swarthmore College, 70–71
Chicago School, 229
Chickering and Gamson's Seven Principles for Good Practice in Undergraduate Education, 134–136
cholera epidemic, research done by John Snow, 141–142
Chomiak, Beverly A.
    bio, 256
    "Religious Studies: Exploring Pluralism and Diversity," 250–255
Christian contemporary music (CCM), geography of, 264
Christianity spreading and Roman trade routes map, 43
Christianity spreading map (300-800 CE), 42
CIHB (Canada Inventory of Historic Buildings) database, use of, 238
circumnavigation of the world (1577-1580), xiii
cities, mapping desirability of, 3–4
classroom GIS, illustrating economics with, 97–99
cognitive maps
    of air pollution and acidic deposition, 24
    application of, 20
"cognitive spatial toolkit," relationship to spatial thinking, 28
Colby College environmental studies project, 187–196
    expanding use of GIS within environmental studies, 193–194
    impact of GIS on teaching and learning, 193–193
    iteration of problems in, 188–192
    lessons learned implementing campuswide GIS, 194–196
Cole, F. Russell et al.
    bios, 197
    "Environmental Studies: Interdisciplinary Research on Maine Lakes," 187–196
Cole, Susan W. et al
    bios, 197
    "Environmental Studies: Interdisciplinary Research on Maine Lakes," 187–196

270

collaborative instruction, importance of face-to-face meetings in, 160
Collaboratory for GIS and Mediterranean Archaeology (CGMA)
fall seminar, 158–161
lessons learned, 164
objectives of, 155–156
practicum, 161–164
relationship to MAGIS, 157
College of Wooster, involvement in archaeological project, 161
Colorado River Basin states household income, salinity, and water flow, 106
communes in Paris
1999 census data for, 231
finding digitized maps of, 230
community and ethnic heritage, links between, 262
community development, use of GIS in, 64
community mapping projects, sustainability of, 72–73
community mapping, roots of, 63
community partnerships, importance in community GIS, 70–72
community-based service, benefits of, xi
computerized cartography, implications of, 45
computers, role in information visualizations, 37
concentric zone maps, use in foreign language and sociology, 233
concentric zone theory, testing on American cities, 229–232
congressional districts, equal population in, 131
Connecticut College religious studies project, 250–255
analysis and presentation, 253–254
data collectors, 251–253
learning outcomes, 255
reflections on challenges, 254
connectivity, role spatial patterns, 28
contiguous United States as blank basemap, xv
cooperation among students, encouraging, 135
cost/benefit standpoint, viewing partnerships from, 71
counts versus ratios, impact on meaning, 51
Cranwell, Richard (former Speaker of Virginia House of Delegates), 132–133
crime statistics, investigation in Habitat for Humanity project, 149
critical thinking
applying to visual correlation, 6, 8
definition of, 22
noticing patterns resulting from, 8
practice of, 6
questioning social assumptions, 8–11
cross-cultural engagement, benefits of, xi
Crude Death Rate, meaning in map of falls by state, xvi
cultural pluralism, accomplishment of, 255
curriculum, quantitative reasoning in, 58–59
curve number method, use in geology project, 214
Cuyahoga River watershed hydric soils map, 178

data. See also spatial data
analyzing in Habitat for Humanity project, 146, 148–149
creating maps from, xv
transforming into smaller meaningful difference, 54–55
visualization of, 27
data acquisition costs, minimizing, 72–73
data collection, performing in religious studies project, 251–253
data display, understanding, 2
data graphics, treatment by Edward Tufte, 37–38
data layers
analyzing and validating spatial relationships between, 148
analyzing for Kilimi area vegetation, 119
application of, 24–25
considering manipulation of, 28
displaying Italian cities in, 3
data relationships, revealing by superimposing maps, 44
datasets, exploring in Habitat for Humanity project, 148
DDViewer Java applet, description of, 107
De Blij, Harm (Why Geography Matters), ix
de Certeau, Michel (spatial stories), 39–41
deaths from falls by state map (2002), xvi–xvii
decennial redistricting for justice and power project, 129–137
deer hunting, economic impacts of, 105–106
Dehaene (perception of quantities), 52
Delaware, OH, collaboration with Ohio Wesleyan, 64–68
Delaware County Trails and Green Spaces map, 65
DEM (digital elevation model), use in biology project, 178–179
demographic maps, benefits of, 234
Denison University chemistry and environmental science project, 201–208
expansion to Environmental Chemistry course, 207
lead deposits in pond sediments, 202–206
lessons learned in field, 208
Density of Concerts Performed by Non-secular Christian Musicians map, 265
DePauw University archaeology project, 155–164
fall seminar, 158–161
Internet-based collaboratory, 79
lessons learned, 164
practicum, 161–164
depth of Togus Pond, impact on environmental health, 188
Depth to Peak Lead Concentrations, 204
Dewey, John, on education, xvii
digital data, availability of, 192
digital elevation model (DEM), use in biology project, 178–179
digital orthophoto quarter quad (DOQQ) images, use in biology project, 181
digitized aerial photos, use in environmental studies project, 189–190
discovery learning, 55–56
districting plans in Microsoft PowerPoint, 135
diversity, exploring relative to religious studies, 249–255

DOQQ (digital orthophoto quarter quad) images, use in biology project, 181
DOQQ of sample sites for environmental science project map, 203
dot density map of Boston census data, 2000, 51
Drake, Sir Francis
    circumnavigation of the world (1577–1580), xiii–xiv
    path near Cape Horn, xiv
    voyage depicted over maximum wind speeds, xiv
Drennon, Christine (Trinity University)
    bio, 152
    Habitat for Humanity in San Antonio project, 79
    "Urban Studies: Assessing Neighborhood Change with GIS," 141–151
Drug Arrests Near a HfH Neighborhood (2003), 149

Easley v. Cromartie (minority representation rights), 131
eastern Massachusetts residents, mapping income of, 8–10
ecological threat, susceptibility to, 188
ecologists, goals of, 172
ecology, mapping population ecology with GPS and GIS, 174–175
economic concepts, analysis of, 99–100
economic impact from deer hunters in Southern Tier West Region, 105
economics
    African Railroads mapping project, 99
    class assignments using GIS, 100–103
    course-based student projects, 103–104
    illustrating with classroom GIS, 97–99
    independent student projects and guided research, 104–107
    linking GIS visualization to fieldwork in, 103
    preference for totals relevant to, 106–107
    reflections and lessons learned, 106–107
    slow adoption of GIS in field of, 106–107
    tool and data review, 107–108
    using interactive maps in, 98
economics project (Siena College), 97–108
education dollars spent in U.S., 52
England, cholera epidemic in (1854), 142
Enoree River Basin, studying in geology project, 215–217
Enoree River Basin watersheds map, 214
Environmental Chemistry course, expansion to, 207
environmental economics assignment, 100
environmental health, jeopardization of, 188
environmental issues, linking to chemistry curricula, 207
environmental science project (Denison University), 201–208
    expansion to Environmental Chemistry course, 207
    lead deposits in pond sediments, 202–206
    lessons learned in field, 208
environmental studies project (Colby College), 187–196
    expanding use of GIS within environmental studies, 193–194
    impact of GIS on teaching and learning, 192–193
    iteration of problems in, 188–192
    lessons learned implementing campuswide GIS, 194–196

ergodic literature, GIS maps as, 45
erosion of soil types, susceptibility of, 205
erosion potential map for Togus Pond, 191
erosion risk theme draped over topographic theme map, 193
ethnic heritage and community, links between, 262
ethnic history and context, modeling use of, 86
evidence, using in reasoning, 11–13
Excel Mapmaker showing unemployment by state, 102
expectations, communicating through high expectations, 136
experts, working memory of, 53

FactFinder Web address, 84
faculty members, involvement in community partnerships, 72
Fairfield University foreign language and sociology project, 227–234
    accessing multimedia resources through maps, 228–229
    learning French and sociology through collaboration, 229–232
Fairhead and Leach (Misreading the African Landscape), 113–114, 117, 122
falls by state, map of (2002), xvi
federal lands in U.S. map prepared in ArcView, 101
feedback, giving promptly, 136
Feletti and Boud (problem-based instruction), 143
Fennessy et al
    "Biology: Spatial Investigations of Populations and Landscapes," 171–182
    bios, 184
Fernald, Cameron (mapping "haves" and "have nots" in MA), 53–54
Figures. See also maps
    African country boundaries, 26
    African Railroads map, 99
    Attleboro mobile home parks project, 93–94
    autism in Rhode Island (2003–04), 7
    bathemetry map for Togus Pond, 189
    bathymetry image for environmental science project, 203
    Bloom's Taxonomy of Learning, 20
    Brazilian ancestry in Boston area, 261
    Brazilian Halloween party data collection, 260
    Catchment Landuse from OCAP Database, 206
    Catchment Soils Classified by Slope Class, 206
    CGMA faculty and students in breakout sessions, 161
    CGMA students using laser transit, 160
    Christianity spreading (300–800 CE), 42
    cognitive map of air pollution and acidic deposition, 24
    Colorado River Basin states household income, salinity, and water flow, 106
    concentric zone theory (Ernest W. Burgess), 230
    contiguous United States as blank basemap, xv
    deaths from falls by state, xvi–xvii
    Delaware County Trails and Green Spaces map, 65
    Density of Concerts Performed by Non-secular Christian Musicians, 265
    DePauw University student practicum, 162–163

Depth to Peak Lead Concentrations map, 204
districting plans in Microsoft PowerPoint, 135
DOQQ of sample sites for environmental science
     project, 203
dot density map of Boston census data, 2000, 51
Drake's path near Cape Horn, xiv
Drake's voyage depicted over maximum wind speeds, xiv
Drug Arrests Near a HfH Neighborhood (2003), 149
economic impact from deer hunters in Southern Tier
     West Region, 105
education dollars spent in U.S., 52
Enoree River Basin watersheds, 214
erosion risk theme draped over topographic theme, 193
federal lands in U.S. map prepared in ArcView, 101
French census data for communes (1999), 231
GDP per capital as GIS view, 98
gerrymandering former Speaker of Virginia House of
     Representatives, 133
Gilder Creek watershed spatial distribution of land uses,
     217
Greencastle sacred landscape, 163
Grinnell College Collaborative project (trees and
     conditions), 67
Habitat for Humanity house locations, 147
heritage buildings in Sackville, 1750–1914, 241o, 242,
     244
Hispanic- and English-speaking groups in Roman
     Catholic community, 253
Hispanic population distribution from 2000 U.S.
     Census, 252
Hurricane Katrina explored through GIS and mapping,
     22
hydric soils in Cuyahoga River watershed, 178
hydrologic distance to stream channel calculation, 179
income disparity in greater Boston, 54
income distribution from 2000 U.S. Census, 252
Japanese auto manufacturing in U.S., 99–100
Kilimi and Kissidougou, West Africa, 115
Kilimi forests organized by size, 120
Kilimi graphs of mean monthly precipitation and
     temperature, 117
Kilimi landscape forest patches, 119
Kilimi villages, 123
Kilimi villages relative to distribution and variation of
     vegetation formations, 121
Kolenten River and Kiilmi basemap features, 118
lakes in central Maine, 187
Leach's adult storm petrel, 175
MAGIS data entry and bibliography pages, 159
MAGIS database search page, 158
MAGIS spatial search for archaeology project, 157
Majority Vote in Each County (U.S.), 27
Mandelbrot set, 38
Mapping Campus-Community Collaborations
     conference (2004), 66
Massachusetts gerrymander of 1812, 130
Massachusetts towns used in Mapping Minorities in
     U.S. course, 90–91
meaning attributed to spatial relations, 2

Microsoft Excel Mapmaker showing unemployment by
     state, 102
minority populations over 30% (U.S.), 86
Mississippi Gulf Coast minority populations above 30%
     of total, 88
Mississippi Gulf Coast minority totals, 87
New Orleans incomes versus home incomes, 103
North America New Map (Jonathan Carver, 1781), 5
Ohio Wesleyan collaborative trails project news article,
     64
Paris metropolitan area showing rings of decreasing
     population density, 233
patriot grave (Otterbein College project), 68
per capita income of black residents in eastern
     Massachusetts, 9–10
PES (Problems in Environmental Science) students, 193
petrel path study area on Kent Island map, 176
polka music versus residents of Polish ancestry, 263
pond bottom interpolations, 205
Pond Depth 1970 map, 204
pond photo for environmental science project, 204
raster maps representing sidewalk density, 57
Religious Communities, New London, CT, 251
Roman Empire roads and waterways, 42
Roman trade routes and spread of Christianity, 43
roof types in Sackville, New Brunswick, 243
Schools in HfH Neighborhoods map, 148
sediment core extracted from pond, 202
Sir Frances Drake's circumnavigation of the world
     (1577–1580), xiii
Smith College locations, 3
Snow's cholera map, 142
spatial distributions for watersheds, 215
subset of spreadsheet showing number of deaths from
     falls, xvi
sulfate deposition amounts, 25
tax parcels for Saratoga Springs (1997), 105
Togus Pond buffer evaluation map, 189
Togus Pond erosion potential map, 191
Togus Pond land-use patterns, 190
Togus Pond septic suitability, 192
Toltec cartographic history, 40
tree species for forest structure analysis, 174
Trimble Pathfinder in use, 176
U.S. Census data, Boston, 2000, 50–51
U.S. presidential election results (2004), 4
Virginia House District 74 (the "hatchet"), 132
wetland restoration potential site suitability, 180
wetland sampling point in Cuyahoga River watershed,
     181
Whitechapel district (Charles Booth, 1891), 12
financial support, leveraging for community mapping
     projects, 72
findings
     communicating for Habitat for Humanity urban studies
          project, 150
     communicating graphically, 145

Firmage, David H. et al.
   bios, 197
   "Environmental Studies: Interdisciplinary Research on
      Maine Lakes," 187–196
FLAC (Foreign Language Study Across the Curriculum)
   program, 227–228
Fletcher, Pamela (art historian), 223
Flynn, John J., documentation of musicians' birthplaces
   by, 266
foreign language and sociology project (Fairfield
   University), 227–234
   accessing multimedia resources through maps, 228–229
   learning French and sociology through collaboration,
      229–232
Foreign Language Study Across the Curriculum (FLAC)
   program, 227–228
forest, challenges to working in, 175–176
forest structure analysis
   establishing sampling points with GIS, 173
   sampling and visualizing data, 172
   technology used in, 173
   visualizing patterns with GIS, 173
Foss, Pedar
   bio, 166
   Internet-based collaboratory developed with Rebecca
      Schindler, 79
   and Rebecca K. Schindler, "Classical Archaeology:
      Building a GIS of the Ancient Mediterranean,"
      155–164
Frames of Mind (Howard Gardner, 1983), 38
France, using as subject in foreign language and sociology
   project, 228–229
French and sociology, learning through collaboration,
   229–232
French census data for communes (1999), 231
Furman University geology project, 213–221
   comparing current and future land-use scenarios,
      217–219
   comparing watersheds with different levels of
      urbanization, 215–217
   incorporating GIS in geosciences, 219–220
   logistics of integrating hydrologic model, 214

Gardner, Howard (Frames of Mind, 1983), 38
GDP per capital as GIS view, 98
Geertz, Clifford (thick description), 41
geographic information systems (GIS). See GIS
   (geographic information systems)
geographic space, personal interpretations of, 20
geography, role in Hurricane Katrina disaster, 20–22
geology project (Furman University), 213–221
   comparing current and future land-use scenarios,
      217–219
   comparing watersheds with different levels of
      urbanization, 215–217
   incorporating GIS in geosciences, 219–220
   logistics of integrating hydrologic model, 214
geosciences, incorporating GIS in, 219–220
gerrymander of Massachusetts (1812), 130

gerrymandering
   connotation of, 132
   significance of, 130–131
Gestalt school of psychology, formation of, 2
Gilbert, Melissa Kesler
   bio, 75
   and John B. Krygier, "Mapping campus-community
      collaborations," 63–73
Gilder Creek watershed, current and future land-use
   scenarios in, 217–219
Gilder Creek watershed spatial distribution of land uses
   map, 217
Gillespie, Dizzy, 265
GIS (geographic information systems)
   accelerating spatial thinking with, 20
   in community development, 64
   creative potential of, 14–15
   establishing sampling points with, 173
   explanation of, xiii
   extending prior knowledge with, 28–29
   as facilitating tool, xi
   illustrating economics with, 97–99
   impact on teaching and learning, 192–193
   incorporating in geosciences, 219–220
   as information visualizations, 36–37
   institutionalization of, 73
   interactive nature of, 13
   and Latin American music, 259–262
   learning through skills and tasks, 145
   lessons learned from campuswide implementation of,
      194–196
   mapping islands and population ecology with, 174–175
   in natural sciences, 167–169
   obstacles to application as pedagogical tool, 30
   potential of, 6
   promoting active learning with, 194
   promoting IDS (interdisciplinary studies) with,
      194–195
   role in building geographic awareness, 21
   role in teaching skills, x
   in service learning, 69–70
   teaching to first-time students in Habitat for Humanity
      project, 144–145
   use in redistricting process, 132
   using in music courses, 262–266
   visualizing patterns with, 173
   worldwide domination of, 63
GIS analysis, performing in historical geography project,
   240
GIS and heritage, teaching, 245
GIS basemaps, creating maps from, xv
GIS component, inclusion in religious studies project, 250
GIS maps. See also maps
   conceptualizing, 23–24
   as ergodic literature, 45
   as sites for narratives, 36–39
GIS model, developing for biology project, 177–179
GIS space, navigating, 45
GIS user interface, considering as calculator, 56–57

GIS-based technologies, market for, 63
Glazer, Jenny (biology project, Kenyon College), 175
global awareness
    developing, 30
    importance in twenty-first century, 19–22
global warming, debates about, 14
"globalization" map, 35
Goldfield, Joel
    bio, 235
    and Kurt Schlichting, "Foreign Language and Sociology: Exploring French Society and Culture," 227–234
Good Practice in Undergraduate Education, 134–136
GPS, mapping islands and population ecology with, 174–175
GPS receivers, popularity of, 167–168
Grady, John
    analysis of U.S. Census data, 78–79
    bio, 95
    "Sociology: Surprise and Discovery Exploring Social Diversity, 83–94
graphics
    treatment by Edward Tufte, 37–38
    versus visualizations, 37
green infrastructure, developing for watershed, 179–181
Greencastle sacred landscape, 163
Grinnell College parks project, 66–67
"ground truth," conceptualizing, 23–24
Guinea savanna ecology in Sierra Leone, mapping, 113–124
guitar, mapping evolution of, 266
Gumprecht, Blake, documentation of musicians' birthplaces by, 266

Habitat for Humanity (HfH) urban studies project, 143–150
Habitat for Humanity house locations maps, 147
Hecht, Brent (geography of Christian contemporary music), 264
Heithaus et al
    "Biology: Spatial Investigations of Populations and Landscapes," 171–182
    bios, 184
heritage and GIS, teaching, 245
heritage buildings in Sackville, 1750–1914 (maps), 241–242, 244
heritage planning
    analytical tool for, 240–243
    contributions to, 243–245
heritage sites, preserving, 237–245
HfH (Habitat for Humanity) urban studies project, 143–150
higher education, critical outcomes of, x
Hispanic- and English-speaking groups in Roman Catholic community map, 253
Hispanic heritage of New London residents, significance of, 250–251
Hispanic population distribution from 2000 U.S. Census map, 252

Hispanic residents in eastern Massachusetts, mapping income of, 10
Hispanics, mapping concentrations in Southwest, 85
historical geography project (Mount Allison University), 237–245
    analytical tool for heritage planning, 240–243
    community outcomes, 243–245
    database development and GIS analysis, 240
    in the field, 239
    further reading related to, 246
    historical source material for, 238–239
    integrating historical geospatial data, 245
historical geography, visualizing in maps, 35
historical geospatial data, integrating, 245
House of Representatives, reapportionment of, 130
humanities course, goal of, 223
Hurricane Katrina (2005), 20–22, 30
hydric soils in Cuyahoga River watershed map, 178
hydrologic distance to stream channel calculation map, 179
hydrologic model
    applications for, 220
    integrating in geology project, 214
hypertext, impact on narrative, 42
hypertext environments, reading text in, 44–45

ICC (intercampus collaborative) teaching, effectiveness of, 155
ICC seminar, design of, 158
IDS (interdisciplinary studies)
    approaches toward, 141–143
    promoting with GIS, 194–195
    reflections on teaching and learning, 150–151
IGN (Institut Géographique National), role in foreign language and sociology project, 230
images, syntax of, 41
income, comparing between home communities and New Orleans, 102–103
income distribution from 2000 U.S. Census map, 252
inferences, making, 29–30
information, choosing display of, xvi
information visualizations
    emergence of, 37
    GIS as, 36–37
INSEE (Institut National de la Statistique et des Etudes Economiques), 230–231
Institut Géographique National (IGN), role in foreign language and sociology project, 230
Institut National de la Statistique et des Etudes Economiques (INSEE), 230–231
intellectual engagement, sustaining power of, 93
intelligences, types of, 38
interaction, role in spatial visualization, 28
interactive maps. See also maps
    use in foreign language and sociology project, 228–229
    using in economics, 98
interactivity, relationship to participatory narrative, 44
intercampus collaborative (ICC) teaching, effectiveness of, 155

interdisciplinary studies (IDS), approaches to, 141–143
Internet GIS, MAGIS as, 157
interpretation, role in problem solving, 23–27
invisible items, visualizing with maps, 2
islands, mapping with GPS and GIS, 174–175
Italian politics seminar, exploration of maps in, 2–4
iterative maps, producing, 39–40

Janelle, Donald G.
    bio, 80
    "Expanding the Social Sciences with Mapping and
        GIS," 77–79
Janvier (translation of symbols), 50–51
Japanese auto manufacturing in U.S., 100
Jones, Robert (GIS applications in economics), 104

Kane, Melinda (environmental economics instructor), 100
Katrina (hurricane), 20–22, 30
Kent Island, population of Leach's storm petrels on,
        174–175
Kenyon College biology project, 171–182
    analyzing forest structure, 172–175
    communicating results, 176–177
    mapping islands and population ecology with GPS and
        GIS, 174–175
    spatial analysis, modeling, and fieldwork for wetland
        restoration, 177–179
    transforming spatial data, 175–177
    working in field and landscape planning, 179–181
K-factors, evaluating in environmental science project, 205
Kilimi, comparing to Kissidougou, 114
Kilimi area
    comparing to Kissidougou, 122
    forest fragments in, 122
    vegetation formations of, 119
    water collection in, 117
    weather in, 117
Kilimi basemap features and Kolenten River, 118
Kilimi forests, distribution of, 122–123
Kilimi forests organized by size map, 120
Kilimi villages map, 123
Kissidougou
    comparing to Kilimi, 114, 122
    forest islands in, 121
    landscape of, 114
Klitgaard, Kent (Wells College Cayuga Lake watershed
        basin project), 104
knowledge
    extending with maps and GIS, 28–29
    synthesizing, 20
Knowles, Anne Kelly (multidimensional geographic
        information), 41
Kolenten River and Kilimi basemap features, 118
Korfmacher, Karl et al
    bios, 210
    "Chemistry and Environmental Science," 201–208

Krygier, John B.
    bio, 75
    and Melissa Kesler Gilbert, "Mapping Campus-
        community Collaborations," 63–73

lakes in central Maine map, 187
land-use changes, measuring effects on water quality,
        213–221
land-use scenarios, comparing, 217–219
language, role of syntax in, 41
language courses, goal of, 223
Le Rouge et le Noir (Henri Beyle Stendhal), 228
Leach and Fairhead (Misreading the African Landscape),
        113–114, 117, 122
Leach's adult storm petrel, 175
Leach's storm petrels, monitoring of, 174
lead, presence in pond (benthic) sediments, 201
learning, encouraging active learning, 136
learning styles, respecting diversity in, 136
legends, questioning, 4
legislative districts, equal population in, 131
L'état, c'est moi (Louis XIV), 233
libraries, role in expanding GIS, 195
Louis XIV (L'etat, c'est moi), 233
L-THIA model, use in geology project, 213–214
Lund, Jennifer J.
    bio, 17, 62
    and Diana Stuart Sinton, "Critical and Creative Visual
        Thinking," 1–15
    and Diana Stuart Sinton, "Thinking with Maps," ix–xii
    and Diana Stuart Sinton, "What is GIS?," xiii–xviii
    "GIS and Spatial Thinking in the Arts, Humanities and
        Languages," 223–225
    "Mapping and Quantitative Reasoning," 49–59
    "Musicology: Mapping Music and Musicians," 259–267

Macelester College land-use plans project, 67
MAGIS, relationship to CGMA, 157
MAGIS data entry and bibliography pages, 159
MAGIS database search page, 158
MAGIS spatial search for archaeology project, 157
Maine
    investigation of water quality in Togus Pond, 188
    lakes in central part of map, 187
Majority Vote in Each County (U.S.), 27
Making Connections: Teaching and the Human Brain
        (Caine and Caine), 11
Mandelbrot set, relationship to information visualizations,
        37–38
map creator, altering relationship with map viewer, 45
map legends, questioning, 4
map reading, discussing, 86
Mapmaker showing unemployment by state, 102
mapmakers
    manipulation of information by, xv
    questioning points of views of, 6
mapmaking, introducing in IDS (interdisciplinary studies)
        class, 150
mapping applications, examples of, 97–98

276

Mapping Campus-Community Collaborations conference (2004), 66
mapping conventions, discussing, 86
Mapping Minorities in the U.S. course
    course structure, 84–85
    developing research project in, 89, 92
    engagement and active learning in, 93
    framing question in, 89
    interpreting pattern and variation on maps, 85–90
Mapping Minorities in the U.S. course (Wheaton College), 55–58
maps. See also Figures; GIS maps; interactive maps
    accessing multimedia resources through, 228–229
    building around spatial narrative, 36
    comparing for deaths by state (2002), xvii
    concentric zone maps, 233
    conveying narrative with, 39
    creating from data, xv
    demographic maps, 234
    extending prior knowledge with, 28–29
    "globalization," 35
    improving math skills with, 56–57
    of income for eastern Massachusetts residents, 8–10
    interpreting patterns and variations on, 85–90
    iterative maps, 39–40
    making and reading creatively, 14–15
    making assumptions from, 3–6
    misleading potential of, 4, 6
    modern maps, 41
    multidimensional geographic information in, 41
    overcoming obstacles to understanding with, 54–55
    role in teaching skills, x–xi
    spatial economy of, 41
    stories contained in, xiv
    superimposing for spatial narrative, 44
    supporting goals of undergraduate education with, x–xi
    temporal maps, 39
    of unrelated data variables, 27
    using as layers in Habitat for Humanity project, 145
    visualizing invisible items with, 2
Massachusetts gerrymander of 1812, 130
Massachusetts map, using in Mapping Minorities in the U.S. course, 89
Massachusetts residents, mapping income of, 8–10
Massachusetts towns used in Mapping Minorities in U.S. course, 90–91
math skills, improving with maps, 56–57
mathematics instruction, contextualizing, 59
Mauck et al
    "Biology: Spatial Investigations of Populations and Landscapes," 171–182
    bios, 184
meaning
    communication of, 2
    drive for finding of, 11
    impact of ratios and counts on, 51
Mediterranean, conducting archaeological research in, 156–157
meetings, importance in collaborative instruction, 160

memory, size limit on working memory, 53–55
"mesh of civilizations," ix
microeconomics, course-based student projects, 103–104
Microsoft Excel Mapmaker showing unemployment by state, 102
Middlebury College GIS in social services project, 68
Miller, Shelie et al
    bios, 210
    "Chemistry and Environmental Science," 201–208
Millsaps students, involvement in archaeological project, 161
minorities in U.S. course (Wheaton College), 55–58
    course structure, 84–85
    developing research project in, 89, 92
    engagement and active learning in, 93
    framing question in, 89
    interpreting pattern and variation on maps, 85–90
minority voting rights, 131–133
Misreading the African Landscape (Fairhead and Leach), 113–114
Mississippi Gulf Coast minority totals, mapping, 87
Mitchell, W. J. T. on images in Western culture, 37
mobile homes in Attleboro research project, 93
modern maps, characteristics of, 41
Monmonier, Mark on computerized cartography, 45
Morrow and Morrow (summer math program), 55
Mount Allison University historical geography project, 237–245
    analytical tool for heritage planning, 240–243
    community outcomes, 243–245
    database development and GIS analysis, 240
    in the field, 239
    further reading related to, 246
    historical source material for, 238–239
    integrating historical geospatial data, 245
Mountains of Kong, nonexistence of, 113
multidimensional critical thinking, awareness of, 23
multidimensional geographic information, significance of, 41
multimedia resources, accessing through maps, 228–229
"multiple intelligences," relationship to spatial narrative, 38
Murray, Janet (participatory narrative), 45
music courses, curricular benefits of using GIS in, 262–266
musical diaspora, mapping, 263–264
musicians and musical groups, mapping, 265–266
musicology project (Wheaton College), 259–267
    collecting spatial data for, 260–261
    curricular benefits of using GIS in music courses, 262–266
    GIS and Latin American music, 259–262
    mapping dissemination, 264–265
    mapping evolution of guitar, 266
    mapping musicians and musical groups, 265–266
Muthukrishnan, Suresh
    bio, 222
    "Geology: Long-term Hydrologic Impacts of Land-use Change," 213–221

narrative. See also spatial narrative
    adding context to, xiv
    conceptualizing, 39
    conveying with maps, 39
    GIS maps as sites for, 36–39
    impact on hypertext on, 42
    participatory narrative, 44–45
    of social life, 89
    synchronic versus diachronic types of, 41
    in two-dimensional space, 41–42
National Academy report on spatial thinking (2005), 28
National Atlas mapmaker, description of, 107–108
natural sciences, teaching with GIS, 167–169
navigation capabilities of GIS, example of, 228
Nelson, Peter (economic geographer), 99–100
New Brunswick, historical geography project conducted
    in, 237–245
New London, religious studies project conducted in,
    250–255
New Orleans (Hurricane Katrina, 2005), 20–22, 30
New Orleans incomes versus home incomes, 103
nitrogen increase, origin of, 216
nonpoint-source (NPS) pollution
    explanation of, 213
    loading values for land-use scenarios, 213, 219
North America New Map (Jonathan Carver, 1781), 5–6
Notes
    "Anthropology: Mapping Guinea Savanna Ecology in
        Sierra Leone," 125
    "Biology: Spatial Investigations of Populations and
        Landscapes," 182
    "Chemistry and Environmental Science," 209
    "Classical Archaeology: Building a GIS of the Ancient
        Mediterranean," 165–166
    "Economics: Exploring Spatial Patterns from the Global
        to the Local," 108–109
    "Environmental Studies: Interdisciplinary Research on
        Maine Lakes," 196–197
    "Expanding the Social Sciences with Mapping and
        GIS," 80
    "GIS and Spatial Thinking in the Arts, Humanities and
        Languages," 225, 234
    "Mapping and Quantitative Reasoning," 60
    "Mapping Campus-Community Collaborations," 74
    "Musicology: Mapping Music and Musicians," 267
    "Political Science: Redistricting for Justice and Power,"
        138
    "Religious Studies: Exploring Pluralism and Diversity,"
        256
    "Sociology: Surprise and Discovery Exploring Social
        Diversity," 95
    "Teaching the Natural Sciences with GIS," 169
NPS (nonpoint-source) pollution
    explanation of, 213
    loading values for land-use scenarios, 216, 219
number sense, utilizing in map reading, 52–53

Nyerges, A. Endre, 79
    "Anthropology: Mapping Guinea Savanna Ecology in
        Sierra Leone," 113–124
    bio, 126
Nyhus, Philip J. et al
    bios, 198
    "Environmental Studies: Interdisciplinary Research on
        Maine Lakes," 187–196

Ohio Rapid Assessment Method (ORAM), use in biology
    project, 180
Ohio Wesleyan University (OWU) collaborative trails
    project, 64–68
    success of, 72
"one-person, one-vote" rule
    constraints of, 132
    significance of, 131
online mapping systems, using in economics, 100–101
ORAM (Ohio Rapid Assessment Method), use in biology
    project, 180
Otterbein College mapping Ohio's revolutionary patriots
    project, 68
OWU (Ohio Wesleyan University) collaborative trails
    project, 64–68
    success of, 72

Paris
    1999 census data for communes in, 231
    finding maps of communes in, 230
    as subject in foreign language and sociology project,
        228–229
Paris metropolitan area showing rings of decreasing
    population density, 233
Park, Robert (Chicago School), 229
participatory GIS, roots of, 63
participatory narrative, explanation of, 44–45
partnership agreements, considering, 71
partnerships
    building in community GIS, 70–72
    success of, 71
    sustainability of, 71
    viewing from cost/benefit standpoint, 71
patterns
    human inclination toward, 15
    identifying within spatial narrative, 43
    interpreting on maps, 85–90
    noticing in critical thinking, 8
    stories encouraged by, 11
    visualizing with GIS, 173
pedagogical tool, impediments to GIS as, 30
per capita income, mapping for black residents of Quincy,
    Massachusetts, 8–10
perceptions, questioning, 8
PES (Problems in Environmental Science) course,
    187–188
PES students (photo), 193
Peterson, Ken
    course-based student projects in economics, 103–104
    use of "live" GIS presentations by, 98

petrel path study area on Kent Island map, 176
petrels, monitoring of, 174
Phillips, Raymond B. et al
    bios, 198
    "Environmental Studies: Interdisciplinary Research on
        Maine Lakes," 187–196
phosphorus levels, predicting for Togus Pond, 191
place
    active and passive exploration, 21
    studying influence of, ix–x
Plebüch, Tobias (musicology professor), 264
pluralism, exploring relative to religious studies, 249–255
points of view, changing, 6, 8
point-source (PS) pollution, explanation of, 213
political concepts in Italy, depicting in maps, 3
political science project
    administering course for, 134–136
    applying Chickering and Gamson's Seven Principles
        for Good Practice in Undergraduate Education to,
        134–136
    census and decennial reapportionment, 130–131
    challenges faced in, 137
    course setting, 133
    legal and constitutional constraints, 131–133
    student projects, 134
political science project (Washington and Lee University),
    129–137
polka music versus residents of Polish ancestry map, 263
pollution
    impact on streams, 217
    point-source versus nonpoint-source types of, 213
Pond Depth 1970 map, 204
pond sediments, presence of lead in, 201
population ecology, mapping with GPS and GIS, 174–175
poverty, relationship to autism in Rhode Island school
    districts, 6–8
PPGIS (public participation GIS), historical geography
    project as, 244–245
practice, importance of, 13–14
practicum for archaeology project
    goals of, 164
    stages of, 162–163
presidential election results in U.S. (2004), 4
problem solving
    importance of, 23
    role of categorization and interpretation in, 23–27
problem-based instruction
    process of, 143
    success of, 150
proximity, role spatial patterns, 28
PS (point-source) pollution, explanation of, 213
public participation GIS (PPGIS), historical geography
    project as, 244–245

quality versus quantity, 13
quantitative literacy, elements of, 58
quantitative reasoning
    in curriculum, 58–59
    discovery learning, 55–56

maps to teach math, 56–57
number sense, 52–53
symbols and sensory input, 50–51
working memory, 53–55
quantities
    contemplating differences between, 51
    perception of, 52
    representing in raster maps, 58
quantity versus quality, 13
questions, forming in Habitat for Humanity urban studies
    project, 144–146
Quincy, Massachusetts
    assumptions about, 10–11
    per capita income of black residents in, 8–10

racial and ethnic minorities project (Wheaton College),
    84–94
racial and ethnic minority groups, categories of, 85
Ramos, Brigitte et al
    bios, 210
    "Chemistry and Environmental Science," 201–208
rapid assessment method, use in biology project, 180
raster maps, using, 57–58
raster surfaces, interpolation in environmental science
    project, 203
rates of change versus ratios, 51
ratios
    versus counts, 51
    versus rates of change, 51
reapportionment process, occurrence of, 130
reasoning, using evidence in, 11–13
"Red" states, interpretation of, 4
redistricting for justice and power project, 129–137
    administering course for, 134–136
    census and decennial reapportionment, 130–131
    challenges faced in, 137
    course setting, 133
    legal and constitutional constraints, 131–133
    student projects, 134
references, 109
    "About That G in GIS," 32–33
    "Anthropology: Mapping Guinea Savanna Ecology in
        Sierra Leone," 125–126
    "Biology: Spatial Investigations of Populations and
        Landscapes," 182–183
    "Chemistry and Environmental Science," 209–210
    "Classical Archaeology: Building a GIS of the Ancient
        Mediterranean," 166
    "Critical and Creative Visual Thinking," 15–16
    "Environmental Studies: Interdisciplinary Research on
        Maine Lakes," 197
    "Expanding the Social Sciences with Mapping and
        GIS," 80
    "Finding Narratives of Time and Space," 46–47
    "Geology: Long-term Hydrologic Impacts of Land-use
        Change," 221–222
    "GIS and Spatial Thinking in the Arts, Humanities and
        Languages," 234–235
    "Mapping and Quantitative Reasoning," 60–62

"Mapping Campus-Community Collaborations," 74–75
"Musicology: Mapping Music and Musicians," 267–268
"Political Science: Redistricting for Justice and Power," 138
"Sociology: Surprise and Discovery Exploring Social Diversity," 95
"Thinking with Maps" (Lund and Sinton), xii
"Urban Studies: Assessing Neighborhood Change with GIS," 152
"What is GIS?" (Sinton and Lund), xviii
reflective process, importance in service learning, 70
relational cartography, definition of, 44
Religious Communities, New London, CT map, 251
religious studies, pluralistic nature of, 250
religious studies project (Connecticut College), 250–255
  analysis and presentation, 253–254
  data collectors, 251–253
  learning outcomes, 255
  reflections on challenges, 254
Reynolds v. Sims (one-person, one-vote rule), 131
Rhode Island, prevalence of autism in school districts in, 6–8
Rhodes students, involvement in archaeological project, 161
Rocky Creek watershed, runoff for, 215–216, 218
Roman Empire roads and waterways map (200 CE), 42
Roman trade routes and spread of Christianity map, 43
roof types in Sackville, New Brunswick, 243
Ross, Glenwood, African railroads mapping project developed by, 99
runoff, predicting relative changes in, 213–214
Rush, Mark
  bio, 138
  "Political Science: Redistricting for Justice and Power," 129–137

Sackville, New Brunswick, historical geography project conducted in, 237–245
Saman, Daniel, 79
  "Anthropology: Mapping Guinea Savanna Ecology in Sierra Leone," 113–124
  bio, 126
sampling points, establishing with GIS, 173
San Antonio, Texas Habitat for Humanity urban studies project, 143–150
Schindler, Rebecca
  bio, 166
  and Pedar W. Foss, "Classical Archaeology: Building a GIS of the Ancient Mediterranean," 155–164
  Internet-based collaboratory developed with Pedar Foss, 79
Schlichting, Kurt
  bio, 235
  and Joel Goldfield, "Foreign Language and Sociology: Exploring French Society and Culture," 227–234
Schools in HfH Neighborhoods map, 148
scientific visualization, integration into coursework, 168
Sears, Ann, documentation of pianists by, 265–266
sediment cores, use in environmental science project, 202

sediment deposition
  calculating volume of, 203, 205
  lead as temporal marker for, 201
sensory input, role in quantitative reasoning, 50–51
service learning
  GIS in, 69–70
  importance of student reflection in, 70
SES (socioeconomic status), role in Hurricane Katrina disaster, 21
sidewalk density, representing in raster maps, 57
Siena College economics project, 97–108
Sinton, Diana Stuart
  bio, 17, 33
  and Jennifer J. Lund, "Critical and Creative Visual Thinking," 1–15
  and Jennifer J. Lund, "Thinking with Maps," ix–xii
  and Jennifer J. Lund, "What is GIS?," xiii–xviii
  and Sarah Witham Bednarz, "About That G in GIS," 19–31
  "Teaching the Natural Sciences with GIS," 167–169
Sir Frances Drake's circumnavigation of the world (1577–1580), xiii
skills and tasks, learning GIS by means of, 145
slave populations, conceptualizing through maps and GIS, 28–29
Smith College student, mapping of campus locations by, 2–3
Snow, John (cholera propagation), 141–142
social assumptions, questioning, 8–11
social life, narrative of, 89
social science instructors, challenges for, 78
social sciences, goal of, 77
social services, using GIS in, 68
socioeconomic status (SES), role in Hurricane Katrina disaster, 21
sociology and foreign language project (Fairfield University), 227–234
  accessing multimedia resources through maps, 228–229
  learning French and sociology through collaboration, 229–232
sociology and French, learning through collaboration, 229–232
sociology project (Wheaton College), 83–94
soil types, erosion susceptibility of, 205
solutions, finding in problem solving, 23
South Carolina, Enoree River Basin geology project conducted in, 215–217
SPACE (Spatial Perspectives for Analysis in Curriculum Enhancement), goal of, 107
space, depicting with tableau spatial stories, 40–41
spatial analysis of Kilimi region, 116–123
spatial data. See also data
  analyzing and evaluating, 27–29
  collecting with music and ethnographic observations, 260–261
  transforming, 175–177
spatial data layers. See data layers
spatial distributions for watersheds, 215
spatial narrative. See also narrative

identifying patterns in, 43
organizing maps around, 36
validating concept of, 43
spatial patterns, explaining in terms of proximity and
    connectivity, 28
Spatial Perspectives for Analysis in Curriculum
    Enhancement (SPACE), goal of, 107
spatial relations
    importance of skills in, 29
    understanding, 2
spatial stories
    building, 45
    categories of, 39
spatial thinking
    acceleration with GIS, 20
    importance of, 19, ix–x
    promoting, 30
    report by National Academy (2005), 28
spatial visualization, role of interaction in, 28
spreadsheet subset showing number of deaths from falls,
    xvi
St. Paul, Minnesota, land-use plans project, 67
Staley, David J.
    bio, 47
    "Finding Narratives of Time and Space," 35–46
states, mapping deaths by falls by, xvi–xvii
status quo map, producing for Habitat for Humanity
    project, 145–146
Stendhal, Henri Beyle (Le Rouge et le Noir), 228
stories, containing in maps and tables, xiv
storm petrels, monitoring of, 174
streams, impact of pollution and nutrients on, 217
student reflection, importance in service learning, 70
student success, impact of symbol translation on, 50
student-faculty contact, encouraging, 134–135
Stump, Roger W. (studies of bebop music), 265
sulfate deposition amounts map, 25
Summerby-Murray, Robert
    bio, 247
    "Historical Geography: Mapping Our Architectural
        Heritage," 237–245
Sunoikisis Project, 155
surveys, administering in religious studies project,
    252–253
Swarthmore College
    collaboration with Chester Consortium for a Creative
        Community, 70–71
    GIS in the curriculum and community project, 68
symbol translation, impact on student success, 50
symbols, role in quantitative reasoning, 50–51
symbols on maps, connotations of, 3–4
synchronic versus diachronic narrative, 41
syntax, relationship to maps, 41

tableau spatial story, explanation of, 39–41
tables, stories contained in, xiv
talents, respecting diversity in, 136
tasks and skills, learning GIS by means of, 145
tax parcels for Saratoga Springs map (1997), 105

teams and team projects
    Carleton College, 66
    Grinnell College, 66–67
    Macelester College, 67
    Middlebury College, 68
    Otterbein College, 68
    Swarthmore College, 68
temporal maps, examples of, 39
text, reading in hypertext environments, 44–45
thick description (Clifford Geertz), 41
thinking globally, ix–x
"Thinking with Maps" (Lund and Sinton), ix–xii
Tierney, Daniel et al
    bios, 198
    "Environmental Studies: Interdisciplinary Research on
        Maine Lakes," 187–196
time, depicting movement through, 39
time on task, emphasizing, 136
Togus Pond buffer evaluation map, 189
Togus Pond erosion potential map, 191
Togus Pond land-use patterns map, 190
Togus Pond, predicted phosphorus levels in, 191
Togus Pond septic suitability, 192
Togus Pond, water quality investigation of, 188
Toltec cartographic history map, 40
tool and data review, 107–108
topology
    appreciating power of, 145
    role in Habitat for Humanity project, 146
tour spatial story, explanation of, 39–40
tree species for forest structure analysis, 174
Trimble Pathfinder in use, 176
Trinity University urban studies project, 141–151
    approaches to interdisciplinary studies, 141–143
    problem-based instruction, 143–150
tsunami in Southeast Asia (2004), 20–22
Tufte, Edward
    use of maps as evidence by, 12
    The Visual Display of Quantitative Information, 37–38
Türk, D.G., eighteenth-century subscription sales by, 264

undergraduate education
    good practices in, 134–136
    using maps to support goals of, x–xi
undergraduates, discovery learning experienced by, 55
understanding, building blocks of, 2
unemployment by state, showing with Microsoft Excel
    Mapmaker, 102
unemployment map (BLS), description of, 108
United States as blank basemap, xv
urban development, relationship to increased nutrient and
    pollution loads, 216–217
urban studies project (Trinity University), 141–151
U.S. Census data, Boston, 2000, 50
U.S. presidential election results (2004), 4
USGS (U.S. Geological Survey) digital aerial photos, use
    in biology project, 181

variables, analyzing relationships between, 8
variations, interpreting on maps, 85–90
Victorian London mapping project, 11–13, 224–225
Virginia House District 74 (the "hatchet"), 132
visual comprehension, relationship to critical thinking, 8
The Visual Display of Quantitative Information (Edward Tufte), 37–38
visualizations
    definition of, 36
    versus graphics, 37
    types of, 37

Washington and Lee University project
    GIS in redistricting, 79
    redistricting for justice and power, 129–137
water quality, measuring effects of land-use changes on, 213–221
watersheds
    comparing with different levels of urbanization, 215–217
    developing green infrastructure for, 179–181
    investigation of Togus Pond (Augusta, Maine), 188
    runoff for Upper Enoree, Rocky Creek, and Brushy Creek, 216, 218
wealth, relationship to autism in Rhode Island school districts, 6–8
Web sites
    FactFinder, 84
    IGN (Institut Géographique National), 230
    INSEE (Institut National de la Statistique et des Etudes Economiques), 230
Wellesley College, requirements for quantitative literacy, 59
Wells College, Cayuga Lake watershed basin project, 104
West African reality, misrepresentation of, 113–114
wetland restoration potential site suitability map, 180
wetland restoration, spatial analysis, modeling, and fieldwork for, 177–179
wetland sampling point in Cuyahoga River watershed map, 181
Wheaton College musicology project, 259–267
    collecting spatial data for, 260–261
    curricular benefits of using GIS in music courses, 262–266
    GIS and Latin American music, 259–262
    mapping dissemination, 264–265
    mapping evolution of guitar, 266
    mapping musicians and musical groups, 265–266
Wheaton College sociology project, 83–94
    course structure, 84–85
    developing, 89, 92
    engagement and active learning in, 93
    framing question in, 89
    interpreting pattern and variation on maps, 85–90
    Mapping Minorities in the United States course, 55–58
    U.S. Census data analysis, 79

Whitaker, Laura B.
    "Anthropology: Mapping Guinea Savanna Ecology in Sierra Leone," 113–124
    bio, 126
white poverty, mapping, 93
white residents in eastern Massachusetts, mapping income of, 10
Whitechapel district (1881) mapping project, 11–13
Whitechapel district (Charles Booth, 1891), 12
Why Geography Matters (Harm De Blij), ix
Wiggins, Thomas Greene, 265–266
working memory, size limit on, 53–55
world circumnavigation (1577-1580), xiii

Yeterian, Edward H. et al
    bios, 198
    "Environmental Studies: Interdisciplinary Research on Maine Lakes," 187–196

zooming technique, relationship to navigation capabilities of GIS, 228

Topical cross-reference

CHAPTER TITLES	Earth and water	Plants and animals	People and census	Human geography and built environment	Cultural awareness	Time and change over time	Community engagement and policy change	Field work	Spatial scale	Inter- and multi-disciplines	Instructional goals and approaches	Locale
What is GIS?	Cape Horn currents		Mortality and morbidity									
Thinking with maps											Critical outcomes of higher education	
Critical and creative visual thinking	Global warming		Autism prevalence; U.S. census data; Italian and U.S. election results; 1891 London census	Perceptions of safe vs. unsafe and known vs. unknown areas	Quality of life in Italy	Persistence of party strength in Italian elections	Survey perceptions of safety and identify problem areas		Local — campus and neighborhood; regional — statewide; national — Italy, U.S.		Using images as instruments of reasoning as well as presentation of conclusions; rapid hypothesizing; pattern-finding; gaining expertise through guided practice	Italy; London; North America; Rhode Island; Smith College in Northampton, Massachusetts
About that G in GIS	Hurricane Katrina		Census patterns in contemporary New Orleans; slavery patterns in the historical South	Clash between built environment in New Orleans and natural disasters	Minorities in New Orleans; slavery patterns in the American South					Inherent connections between geography and all other academic disciplines	Spatial thinking integrated into the critical thinking process; inquiry-based learning; problem solving; Blooms Taxonomy of Learning	New Orleans; Africa; American South
Finding narratives of time and space	Dust Bowl			Roman roads and pilgrimage paths	Dissemination of religion						Generative act of mapmaking	
Mapping and quantitative reasoning			Autism prevalence; U.S. census data; economic disparity by race and location		Expenditures on education by state	Widening gap of median town incomes			Local — cities; regional — statewide; national — U.S.		Discovery learning; sensory experience with quantities; building skill and confidence with quantitative reasoning	Massachusetts, Rhode Island, U.S.
Campus–community collaborations	Watershed pollution	Invasive plants; trees in parks	Urban poverty	Land-use goals; urban development; historical graves; social services; "precious places"			Map community assets; recreational trail; invasive plants; land-use plans; map history; map pollution		Local — campus and nearby town	Urban planning; geography; biology; social services; history	Service-learning	U.S.

SOCIAL SCIENCES

CHAPTER TITLES	Earth and water	Plants and animals	People and census	Human geography and built environment	Cultural awareness	Time and change over time	Community engagement and policy change	Field work	Spatial scale	Inter- and multi-disciplines	Instructional goals and approaches	Locale
Sociology: Surprise and discovery exploring social diversity			Minorities in Massachusetts and U.S.; urban; female-headed households; housing value; per capita income; urban poverty	Mobile homes	Minorities in Massachusetts and U.S.			Observation and documentation of urban areas	Local — cities near campus	Sociology and economics	Hypothesis generation	Massachusetts; U.S.
Economics: Exploring spatial patterns from the global to the local	Sewage, water quality, and housing prices; economic development and water	Deer hunting	UN Human Development Index; unemployment; Income distributions; Population growth and public land; Hedonic price model	African rail corridors; factory locations; social service sites	Understanding of poverty and social class stereotypes	Changes in unemployment rates over time		Observation of businesses and neighborhoods	Local - citywide projects in New York and Louisiana; regional - U.S. regions; continental - Africa ; global - worldwide patterns	Economics and the environment; economic geography	Deepen understanding of economic structures; Facilitate writing about economic outcomes	Campus locale; Chesapeake Bay; New York State; midwest U.S.; western U.S.; Africa
Anthropology: Mapping Guinea savanna ecology in Sierra Leone	Seasonal streams; topography; water availability; seasonal climate	Forest distribution; vegetation associations; primates	Susu of Sierra Leone	Village housing; agriculture; land use	Landscape representation	Forest growth; settlement patterns	Conservation implications	Ethnography and local ecology	Local — village-level; regional — Western Africa; Guinea savanna	Integration of anthropology and ecology through spatial analysis	Focus on spatial analysis, while learning software skills from self-paced modules and practice	Western Africa Guinea savanna; Kissidougou; Kilimi
Political science: Redistricting for justice and power			Demography and votes	Gerrymandering	Political manipulations		Hypothesize improvement to local voting districts		Statewide — Data covering all of Virginia			Virginia
Urban studies: Assessing neighborhood change with GIS			Characteristics of Habitat for Humanity communities	Education; investment; crime			Assess Habitat for Humanity		Local — City of San Antonio, Texas	Urban studies; sociology; anthropology; political science; economics	Project-based learning format; sequence of teaching GIS skills	San Antonio, Texas
Classical archaeology: Building a GIS of the ancient Mediterranean			Mediterranean basin; social changes; political changes			Historical development; disciplinary development	Archaeological field survey; remote sensing; artifact analysis	Local—field projects in a village or town; Regional/national—data collection for studies around Mediterranean Basin	Classical studies; Mediterranean archaeology; ancient history; geology	Basic cartographic and GIS skills; survey archaeology; research design and implementation; bibliography hunting	Mediterranean Basin

TOPICS

Chapter Titles	Earth and water	Plants and animals	People and census	Human geography and built environment	Cultural awareness	Time and change over time	Community engagement and policy change	Field work	Spatial scale	Inter- and multi-disciplines	Instructional goals and approaches	Locale
NATURAL SCIENCES												
Biology: Spatial investigations of populations and landscapes	Forest floor; wetland restoration; watershed assessment; wetland modeling	Forest diversity; bird nests; wildlife habitat				Change over time in tree species; 50 years of bird nest locations	Participate in conservation planning	Forest and wetlands assessment; bird observations	Local – individual forest stands; bird nesting sites; Regional – countywide wetlands assessments		Field research experience; active learning	New Brunswick, Canada; Ohio
Environmental studies: Interdisciplinary research on Maine lakes	Algal blooms; fertilizer runoff	Algae		Water pollution; soil erosion; residential development		Annual analyses of lakes since 1988	Present data to lake associations and Environmental Protection Agency; create atlas of Maine	Bathymetry; water sampling	Local – individual ponds or lakes	Chemistry; economics; government; zoology	Active learning; civic engagement	Maine
Chemistry and environmental science: Investigating soil erosion and deposition in the lab and field	Lead levels; erosion; sedimentation			Riverbank erosion		Using lead depositions to mark a point in time		Use GPS; water sampling; core samples	Local – individual pond	Chemistry; earth science	Project-based learning; research experience	Ohio
Geology: Long-term hydrologic impacts of land-use change	Floods and runoff; Water quality; Water consumption; Land-use planning; Resource management		Population change	Water quality; Urban development; Industrialization; Smart growth		Forecasting future runoff from modeled growth and development	Interact with community planning groups regarding future development		Local/Regional – watersheds (35–40 km²)		Teach hydrology, land-use planning, and simple GIS modeling	S. Carolina
ARTS, HUMANITIES, AND LANGUAGES			1891 London Census		Art history						Virtual field trip	Whitechapel, London
Foreign languages and sociology: Exploring French society and culture			Population density	Locations of historical and cultural landmarks	Basketball; countryside; historical influence on settlement patterns	Settlement patterns across centuries			Regional – Metropolitan Paris; Continental – European cities	Cultural studies and Sociology	Teach awareness of place; Virtual field trip	France, esp. Paris
Historical geography: Mapping our architectural heritage			Settlement	Historic buildings and architectural characteristics	Architecture	Building inventory; settlement patterns	Preserve tourism	Identification and verification of historic buildings	Local – field work in town near campus	History; Architecture	Research experience	New Brunswick, Canada
Religious studies: Exploring pluralism and diversity			Income; ethnicity; religion; race; language	Locations for church, synagogue, mosque, and other religious communities	Religion	City transition from prosperity to low income	Present to religious organizations	Ethnographic observation	Local – field work in city near campus	Demographics and religion	Integration of research, pedagogy, and service to the community	Campus locale, Connecticut
Musicology: Mapping music and musicians			Diasporas; music preferences and ethnicity; regional purchases and preferences	Ethnic music venues and cultural context	Christian contemporary music; polka	Jim Crow laws and musicians' careers; heritage of opera singers; evolving guitar technologies		Ethnographic observation of cultural context of music venues in nearby city	Local – field work in cities near campus; Regional – patterns and trends within U.S.	Ethnography; demographics and music	Research experience	New York, Boston, Austria, U.S.